WATERMILLS
OF THE LONDON COUNTRYSIDE

*

THEIR PLACE IN ENGLISH LANDSCAPE AND LIFE

VOLUME TWO

Rickford Mill, Worplesdon, Surrey (1931)

WATERMILLS OF THE LONDON COUNTRYSIDE

Their Place in English Landscape & Life

VOLUME TWO

KENNETH C. REID
A·R·I·B·A

With Drawings and Diagrams by the Author of Examples Past and Present

Warnham Mill

CHARLES SKILTON LTD

Made and printed in
Great Britain by
Mackays of Chatham plc
Letchworth, Hertfordshire
and published by
CHARLES SKILTON LTD
Whittingehame House
Haddington, Scotland

ISBN 0284 98806 5

CONTENTS
VOLUME TWO

Ruins of mill near West Drayton

LIST OF ILLUSTRATIONS

MAPS

Chapter 12

THE WEY AND TRIBUTARIES

WHITLEY BROOK

THE Wey basin has several tributaries and a start can be made with the Whitley Brook and with Ockford flour-mill, Godalming, near the railway. It represents the South Mill attached to the manor of Laborne. Since the last war, when the flour milling ceased here, this building has become an office and stores with some additional buildings, all plant having been removed to leave only the structure of *c*1850. The pool fortunately remains and so does the spillway. When visited in 1930 this mill still flourished, though some time around 1900 there had been a substitution of a turbine for a water-wheel.Inside the smell of meal impregnated the air and a layer of flour dust clung to each and every surface. The miller at that time mentioned that he had to take care in regulating the water from the extensive ponds and whenever he allowed more than the usual quantity to pass through the mill or down the spillway he had to inform those at the mill lower down otherwise its overshot wheel would suffer.

On the stone floor a set of stones is coupled to a sieve while on the floor above could be seen the sifter and other apparatus. A bran room and old apartments used as living rooms by previous millers stood at the rear. On the attic floor were the storage bins and the usual mechanism for hoisting the sacks, in other words a more or less standard mid-Victorian type mill. Just down the stream from Ockford mill and close to the railway bridge could once be seen an undershot wheel, a remnant of one of the tanneries which in the last century stood hereabouts.

Below Godalming station can be found Godalming Mill. Even forty years between visits still allows it to be recognized, though one must admit there has been a certain deterioration in interest and surroundings. In those days it produced flour but is now a store and an adjunct to the factory which has arisen alongside and in a sense replaced it. The plant no longer remains and the wide overshot iron water-wheel had been replaced in 1944, so it seems, by a turbine which however efficient is much less eyeworthy. In the main the building was late eighteenth century to early nineteenth century, utilizing considerable portions from the earlier mill. Inside one can see the medieval masonry wall acting both as a dam to the pond, now dried up, and as a retaining wall supporting the side passage. A concrete floor replaces its original and generally the structural condition of the mill is not good and one wonders how long it will survive.

The construction is far from straightforward, especially when one walks along the bin floor, where most of the bins remain and where the roof is somewhat of a puzzle. Parts of the latter have been reboarded and tiled. Some lengths keep a ridge tree. In the case of others, the rafters are halved at their junctions and are housed on to the plates and purlins, thus indicating a rather earlier origin than the rest of the mill would suggest. It may be that the roof and part of the structure was left when the two lower floors were rebuilt from within in the eighteenth century. Also some of these pine beams have sagged six inches over their length, whether from overloading or poor timber may be unclear, but such failing is rarely seen to happen when oak has been used.

THE HASCOMBE STREAM

Between the last stream and the Tillingbourne, there flow three streams whose names seem uncertain. A mention should be made of Vachery near Cranleigh, a fair sized sheet of water. The extent dates from an enlargement made about 1760, to act as a reservoir for the Wey and Arun Canal, but an earlier, smaller predecessor supplied some ironworks which closed down after 1581 when the owners were the Bray family, also owners of the Abinger Hammer iron-mills. This stream, *en route* to the Wey, is joined by two others, one of which comes down from Hascombe. Hascombe Mill, first on the stream, has gone while the next is

Ockford Mill, Godalming (1931)

Tannery Water-wheel, Godalming (1931)

Thorpe Mill, Egham (1932)

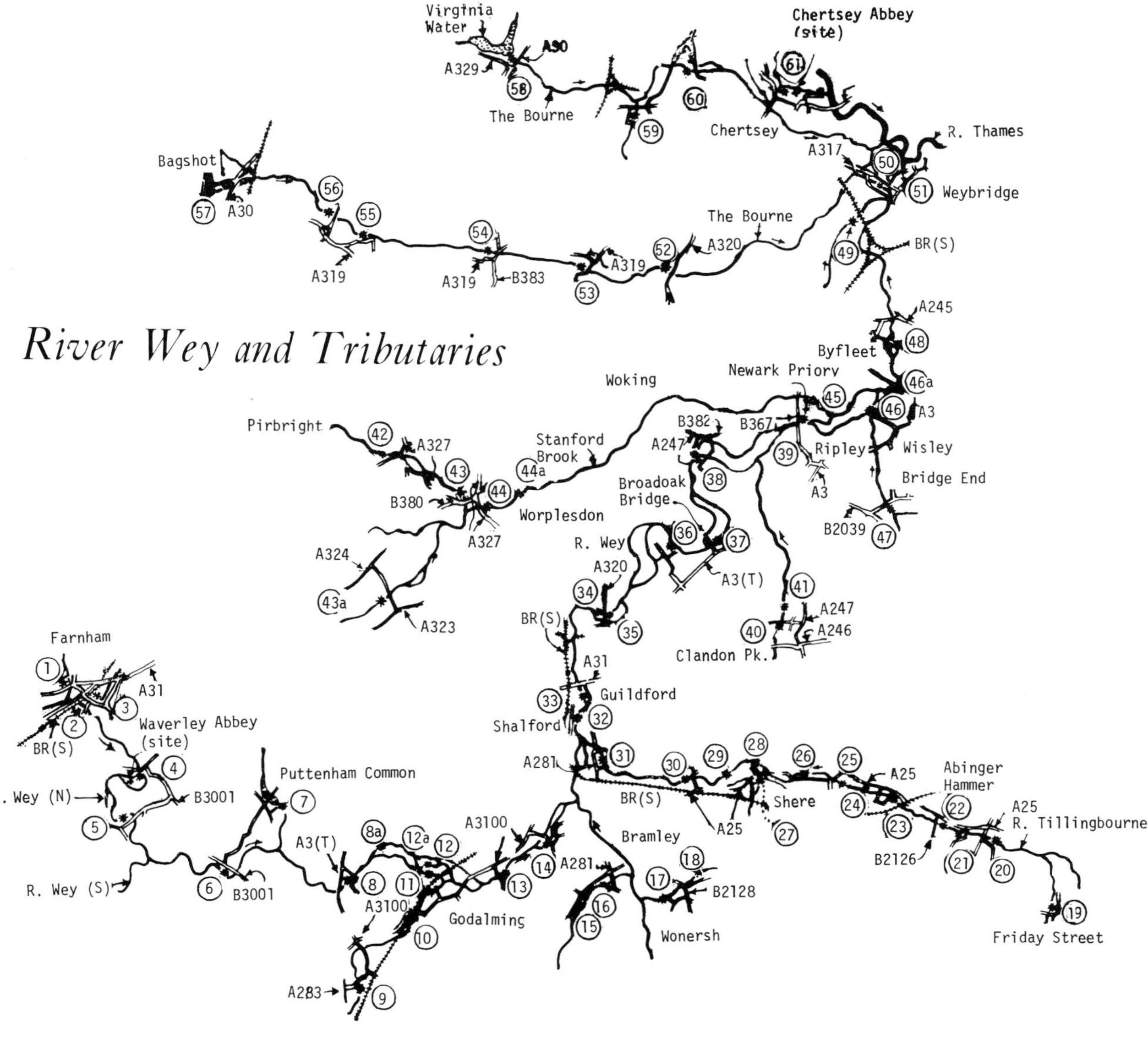

1 *Bourne Mill*
2 *High Mill*
3 *mill site*
4 *Waverley Mill*
5 *Tilford Mill*
6 *Elstead*
7 *Cutt Mill*
8 *Eashing Mill*
8a *mill site*
9 *Whitley Mill*
10 *Ockford Mill*
11 *Godalming, Town Mill*
12 *Salgasson Mill*
12a *Leather Mill*
13 *Catteshall Mill*
14 *Unstead Mill*
15 *Snowdenham Mill*
16 *Bramley Mill*
17 *Wonersh Mill*
18 *Lostiford Mill*
19 *mill site*
20 *Abinger Mill*
21 *Paddington Mill*
22 *Abinger Hammer Mill*
23 *Gomshall Mill*
24 *Netley Mill*
25 *mill site*
26 *Albury Mill*
27 *Upper Mill, Postford*
28 *Postford Mill*
29 *Chilworth Gunpowder-Mills*
30 *Chilworth Paper-Mill*
31 *Pratts Mill*
32 *mill site*
33 *Guildford Mills*
34/35 *Stoke Mills*
36 *Bowers Mill*
37 *mill site*
38 *St Martha's Mill*
39 *Newark Mill*
40/41 *mill sites*
42 *Upper Mill*
43 *Heath Mill*
43a *mill site*
44 *Rickford Mill*
44a *Blanket Mill*
45 *mill site*
46 *Ockham Mill*
46a/47 *mill sites*
48 *Byfleet Mill*
49 *Cox's Lock Mill*
50 *mill site*
51 *Ham Mill*
52 *Dunford Bridge Mill*
53 *Emmetts Mill*
54 *Chobham Mill*
55 *Hook Mill*
56 *Windlesham Mill*
57 *Bagshot Mill*
58 *Harpsford Mill*
59 *Trumps Mill*
60 *Thorpe Mill*
61 *Chertsey Abbey, site of mill*

Wonersh Mill, Surrey (1931)

Bramley Mill, Surrey (1933)

Paddington Mill, Abinger (1931)

Albury Mill

Netley Mill, Shere (1931)

Snowdenham Mill, once known as Snadlam Mill,[1] and, though its working life ceased before the First World War, its presence today is most substantial. Not only do the pond and the gardens of Snowdenham House lend it a special setting but the interior offers much of interest. The drawings should save a description, except to mention that the timber frame would be of the sixteenth century or earlier, with later rebuildings and renewals. It lacks a familiar feature — the lucam — because the grain sacks could be delivered direct to the stone floor.

However, it seems that the dam was raised about 1886 when the wheel and plant were renewed. Prior to this date the service road would have been lower than the present lane. Masonry in the 'sack', or lower, floor would indicate the original level of access. Previous to this reconstruction the wheel may have been undershot in contrast to the present overshot one. The brick upper storey could also date from this rebuilding but the roof, with its rafters being tenoned into the purlin and the latter into the trusses, belongs to much earlier work. Two other features are worth noting: the unusual one of a set of carved spur brackets on the wooden upright shaft, helping to support a cast iron wallower; and the survival *in situ* of two excellent flour dressers.

Bramley Mill no longer works (it did as late as 1934) and the water-wheel on the north side has been removed, as obviously has the remainder of the plant. In such a pleasing situation, with as fine a pond as any in Surrey, the building invited re-use as a dwelling and this it has become. A probable date for the mill building would be 1830, while the attached miller's house was of the late sixteenth century. All of this, of course, now forms one edifice, but the old 'Folly Tower' seems a strange adjunct to the wall of the mill, its purpose being far from evident. Another stream rises below Blackheath, and fills a pond that drives Wonersh Mill. The situation is most worthy and the nineteenth-century fabric (bereft of the plant inside) is well looked after. One must be happy with such a state to the alternative, and more often occurring condition, yet be reluctant, if resigned, to accept this over-tailored tidiness that so much of old Surrey wears today. Before this stream reached the last-mentioned mill, there stood Lostiford Mill. Only the spillway, landscaped into the adjoining garden, remains to indicate this one-time complex of buildings.

RIVER TILLINGBOURNE

This river, like many others, rises below Leith Hill, and one of its head streams at Friday Street filled a 'mill pond', but no mill building has come down, and likewise for Wootton. There was one once, held in 1377 by William Lattimer, and probably giving place to the iron, wire-drawing and gunpowder-mills recorded later, for which tradition points to George Evelyn as the founder. An explosion greatly damaged both his mills and his home, Wootton House, not far away. The stream, for that is what it is, flows through this delectable parkland to the pond serving Cranes, or Abinger, Mill. The pond has virtually disappeared and so has the seventeenth-century watermill, leaving only the lower stone walls to mark its position.[2] Next downstream, the remains of another fine pond point to Paddington Mill, a small but not unattractive Victorian building. The wheel has gone, leaving the axle, otherwise the picture here presented in the early thirties is readily recognizable.

Paddington Bray, the name of the chief manor in these parts, belonged in 1305 to Adam Gurdon who then had a mill. This would have been the one next below the stream to Paddington Mill and was called Abinger Hammer Mill, which gave the name to this little village. In the sixteenth century it belonged to Ann and Owen Bray who, in 1556, sold it to Thomas Elrington, a Willsden man who received in 1560 a special licence to build an iron mill. Though the permission was to last fifteen years it did not prevent a petition being got up by the local gentry in 1566 against his 'late erected yron mylle'. However, the petition failed, as in 1574 Edward Elrington of Harlsden inherited the mill from his brother Thomas and apparently it kept at work until the early nineteenth century. It is difficult to point to the exact site and though Rocque[3] shows it on the west side of the Holmbury Road no relic is now visible there.

Gomshall and Shere Mills were mentioned in the thirteenth and fourteenth century, both belonging to Netley Manor, previously the property of the De Es family. The owner in 1233, Sir Mathias Besille, bestowed the manor upon Netley Abbey in Hampshire. The name has clung to the example near Shere. The De Burgh family had obtained a portion of Shere Ebor manor and two members drew rent from Netley Mill: Richard, Earl of Ulster, in 1297 and his son, William, in 1344. Gomshall Mill ceased grinding corn after the last war. Since the early thirties, its surroundings have lost their unspoilt note, though at present care is evident. The stepping stones that previously provided a crossing over the shallow ford have been replaced by a bridge, while a car park occupies the tea gardens once overlooking the tail stream. Fortunately the structure itself has been saved by becoming a restaurant.

The mill-house adjoining on the north now accommodates an antique showroom, the mill itself and former stores being used for the restaurant and kitchen.[4] A certain number of removals and additions have become evident. Fortunately the oak has been left natural, not blacked over, though the fibreboard panel infilling and terrazzo tile floors are somewhat out of character. The extension built in front of the wheels in the early nineteenth century now provides a viewing platform for the water-wheel which can be seen through the glazed panel. The second water-wheel and plant have been removed, probably just before the war, and it can be noticed that the overflow spillway stood between the two wheels, an arrangement which once could be seen externally at the mill at Church Cobham.

Of the second plant, in iron, some of it has survived: a partly broken cast iron pit-wheel driving a cog, or wallower, on a line (or lay) shaft, which powered two sets of stones, and housed within the altered structure of the hurst which is obviously much older than this particular late nineteenth-century machinery. Concerning the structure itself, while enjoying a meal, it is pleasant, if difficult, to try to unravel how this watermill acquired its final shape. The rubble masonry walls could easily be of the fifteenth century. Most of these remain, while the plate carrying the timber

Gomshall Mill (1931)

superstructure can be recognized, and also some of the upright posts, but when the projecting wing was built in the late sixteenth or early seventeenth century to house bins, the framing suffered all sorts of alterations and additions. The original timbering can still be seen above the missing water-wheel. Altogether Gomshall Mill as it now stands is well worth a visit.

Netley Mill has not ground meal for a long time, being part of the local waterworks, as the rather unpleasantly obtrusive building erected nearby confirms; not that the local scene in this particular area could suffer any further desecration. This mill building uses local stone of as many hues as the Vale of Holmsdale's characteristic building materials permit, but the diminutive tower attached to the corner is a puzzling feature. It recalls a similar tower at Bramley Mill not very far away and this also indicates that, despite the 'Gothic' windows, the building is of much the same date, i.e. early nineteenth century.

The Tillingbourne, after filling the sawmill pond, passes through the village of Shere and then enters Albury Park. A corn-mill once standing here survived the destroyed village and in 1814 could be described as the chief paper-mill in Surrey. The next, Albury Mill, alongside the village stream, worked as a pumping station for the waterworks. That was thirty years ago, today it functions as offices and laboratory for a large concern. An obvious change is that, since then, pool and leat have been filled in, and paved over, though the railing can still be recognized. The shell of the watermill, altered and dolled up with the currently fashionable detail which would be as expected, seems to appeal today more than it did when the drawing was made.

Postford Flour Mill looks like a factory, though in a pleasing setting, with nothing apart from the overflow to indicate a long-established site. Another watermill (its shell still remaining) stands by a pond a little distance southward up the hill, but a turbine now replaces the former water-wheel. This mill, which may represent a Domesday site, processed flock and not corn up to 1929 and obtained its power from a separate tributary. The tail from the mill flows in a man-made channel along the old parish boundary, joining the Tillingbourne just below Chilworth Mill. Beforehand a branch flows to the site of the gunpowder-mills.

The Tillingbourne flows through the thickly wooded valley that Cobbett calls his 'little romancy vale',[5] a title hardly applicable today. Hereabouts, happily hidden by trees, stood the ugly remains of St Martha's works, the gunpowder-mills which must have been closed after the First World War. At some point below the bridge stood the powder-mill, famous since Elizabeth's day, i.e. Chilworth Mill, which probably provided power for the works. Tradition makes the date of the establishment, *c.*1570, a result of the licence granting permission to the East India Company to make all the powder it required for its own use, but this could hardly be because the Company was not formed until 1599.

Aubrey, in mentioning this little group, says that the Tillingbourne had sixteen or eighteen of these gunpowder-mills[6] and that 'tis a little common wealth of powder makers who are as black as negroes'. Evidently explosions occurred often and on one occasion the force was so great that it shot a piece of timber through a cottage wall and knocked off a poor woman's head while she sat spinning.

Further downstream, at the corner of Penny Lane, can be found the mill-race of Chilworth Paper Mill, burnt down many years ago. This example is claimed to be the gunpowder-mills established *c.* 1570, but as corn-grinding and cloth-fulling were carried on here in 1589, confusion has arisen in the tradition. A drawing[7] of the mill shows it to have been timber built on a brick ground storey, probably of the seventeenth century with a later and prominent chimney. It was converted in 1704 into a paper-mill and over a hundred years later it incited Cobbett's righteous indignation.

> This valley which seems to have been created by a bountiful providence, as one of the choicest retreats of man, which seems formed for a scene of innocence and happiness, has been by ungrateful man, so perverted as to make it instrumental in effecting two of the most damnable of purposes; in carrying into execution two of the most damnable inventions that ever sprang from the minds of man under the influence of the devil! namely the making of *gunpowder* and of *banknotes*. Here in this tranquil spot, where the nightingales are to be heard earlier and later in the year than in any other part of England; where the first bursting of the bud is seen in Spring; where no rigour of seasons can ever be felt; where everything seems formed for precluding the very thought of wickedness; here has the devil fixed on as one of the seats of his giant manufactory; and perverse and ungrateful man not only lends his aid, but lends it cheerfully! As to the gunpowder, indeed, we might get over that. In some cases that may be innocently, and, when it sends the lead at the hordes that support a tyrant, meritoriously employed. The alders and willows, therefore, one can see, without so much regret, turned into powder by the waters of this valley, but *the Banknotes*! To think that the springs which God has commended to flow from the sides of these happy hills, for the comfort and the delight of man; to think that these springs should be perverted into means of spreading misery over a whole nation; and that, too, under the base and hypocritical pretence of promoting its *credit* and maintaining its *honour* and its *faith*! There was one circumstance, indeed, that served to mitigate the melancholy excited by these reflections; namely that a part of these springs have, at times, assisted in turning rags into Registers!

The last watermill on the Tillingbourne lies behind Shalford village street. It had long been disused and in 1932 a fateful 'For Sale' board appeared, boding ill for its future. Fortunately the National Trust took it over. Its gift to that body with the necessary endowment was made possible by the good offices and generosity of persons then unknown who kept their anonymity under the name 'Ferguson's Gang'. This rather amusing but effective affair can best be told from the pages of *The Times*:

> On January 13 [1933] . . . a taxicab stopped outside 7 Buckingham Palace Gardens [National Trust offices]. A lady, heavily masked, got out and announced herself to the commissionaire as 'Red Biddy of Ferguson's Gang'. She was bent with the weight of a heavy bag and desired to see the secretary, with whom, without lifting her mask, she deposited £100 in silver as the first instalment of the Shalford Mill endowment. She left, as she had come, unrecognized . . .

Such a sum belongs most eloquently to pre-inflation days.

A path leads to the mill, and its special feature, that curious 'porched' lucam, probably one of the late seventeenth-century additions. Laden carts would have come under the upper storey and the wheat hoisted up into the loft and thence to the storage bins. The age of building could start with the late sixteenth century, followed by many alterations and replacements such as the brick lower storey pierced by three narrow arches. Inside a collection of timber framing both in oak and in pine can be examined, also the much repaired windows of the eighteenth century, mostly, and also what is rare, some feathered weatherboarding. The roof structure without ridge-board looks older than the floor and wall construction below and well it may be, being somewhat similar to that at Godalming Mill.

Most of the plant, an early nineteenth-century layout, has not been removed and forms one of the best near London. It comprises a cast iron water-wheel, the upright shaft, wallower, spur-wheel, two stone nuts and two (once three) sets of stones, with a considerable survival of the flour dressing apparatus. The repair of the place is sufficient to preserve it with none of the undue restoration that sometimes happens in such cases. The mill proper serves as a store cum workshop to the dwelling adjoining, a fact much in evidence when exploring the mill's three storeys, and which imparts a pleasing casual air to the place.

Shalford Mill porch (1931)

Chobham Mill, Surrey (1931)

Shalford Mill (1931)

STANFORD BROOK

This stream with some tributaries joins the Wey near Ockham. Henley Park near the hamlet of Normandy has a weir, seemingly the site of a watermill belonging to Chertsey Abbey. About 1322, owing to some failure perhaps, the abbot ordered its removal and replacement by a windmill. Near Pirbright Manor House, on the Hodge Brook, behind a high pond and waterfall is the Pirbright or Upper Mill. This has not worked for many years but it remains, and with the adjoining one-time miller's house, gardens, pool, streams and accompanying barns forms a very attractive composition. The building dates from about 1800 with the remains of a half-timbered truss and the usual bits and pieces of wood from an earlier building. The interior is more interesting than the outside. The remains of the cast iron water-wheel frame can be seen, as can the penstocks controlling the water, worked from a raised platform near the stone floor. The cast iron wallower and wooden spur-wheel are still there. Also remaining is an upright shaft carried with its bearing on a rather large iron bridge and driving a wooden spur-wheel which in turn drove three sets of stones rather closer together than is usual. At the top lies the crown-wheel, cast iron with wooden cogs. The owner voiced the hope that sometime in the future there would be the possibility of renovating this still interesting specimen.

The stream flows down the Common to Heath Mill

Heath Mill, near Woking (1931)

Near Woking (1931)

Town Mill, Guildford (1928)

which also has ceased grinding and a recent conversion has made it into an attractive picture. It is mid to late Victorian, in brick, and resembles in certain details the mill at Ockham a few miles further down on the Wey. The conversion allows the iron water-wheel, partly complete, to be seen, also the pit-wheel, the wallower and the spur-wheel.

The next mill, nearer Worplesdon, is Rickford Mill. In earlier days called Perry Mill it, until recently, still ground corn seed and cake, but one missing element is a magnificent chestnut tree that once overhung the pond. The main building is eighteenth century, with other walls of a much later date, but it no longer works now and the structure has been gutted, presumably ready for conversion into a house of which some new portions can already be seen to have been erected. The water-wheel has gone and the pit has been filled in and it is hoped that the fine wooden spur-wheel, the wooden stone nuts, the iron crown-wheel with its wooden cogs and upright shaft can somehow or other be retained in the new use of the building. The brook flows past Blanket Mill Farm where once stood a mill and another further downstream credited to Pyrford has also disappeared.

WINDLE BROOK

This stream rises in Bagshot Park. It changes its name, first to the Hailbourne, and then to the Bourne, before joining the Wey and Thames. It fills a pond near Bagshot Church, or it did so, and in this case the lade, high-banked, conveying the water to the head race, represented quite a considerable piece of earthwork engineering. The building of the watermill is still there with the miller's house adjoining. A tablet with the date '1817' built into the wall proclaims its age and the mill now functions as a workshop and office for a firm of builders. Rocque does not show it on his map of Surrey, so it may date as the wall tablet says. Further downstream could once be found Windlesham Mill by Broadway Green, and Hook Mill; water channels and spillways only remaining. The stream flowed on to Chobham Mill. Around 1930 this was a pleasant spot, particularly the pool, now shrunken from its former extent. But of the mill only the lower floor and part of a wall with the spillway alongside can be seen. The axle 'arch' can also be recognized and, of course, the mill-race, but the area of the mill itself has now been built on. This example is doubtless the descendant of the 'Hurst Mill' conveyed by John de Hamme to the abbey of Chertsey in the fourteenth century.

Further eastwards the Hailbourne passes under a humped back bridge at what was Emmetts Mill. This building must have ceased grinding corn before the First World War, and sometime in the thirties it was converted to a house and much rebuilt in the change. The stream still tumbles down the spillway and the cast iron wheel frame (less paddles) looks little different in the last thirty years. Downstream near Dunford Bridge, before the Wey is joined, some waterways indicate that here stood a mill of the same name, concerning which records seem silent.

RIVER BOURNE

The Bourne rises in the purlieus of Windsor Great Park and the first watermill seems to have been that at Sunninghill. This was consumed by a fire in 1890 so it is presumed no remains are there today.

The stream ran originally in a valley which at a later date was filled by Virginia Water, and where the Roman road from Staines to Silchester crosses that stream, stood Harpsford Mill. It belonged to the Crown and its profits, which came to £4 in 1250, had sadly diminished by 1315, probably because Old Windsor had declined leaving few customary tenants to support this mill. However, in 1340 the constable to the Castle reported that as the building had fallen into ruins, no rent at all could be obtained from it.

A mile or so downstream stands Trumps Mill, evidently one that represents a mill belonging in 1299 to Henry and Matilda Middleton. One of their descendants sold the manor of Milton, including Trumps Mill, to Richard Fox who had already purchased other lands in the parish. This man, Bishop of Winchester, statesman, soldier, herald and diplomat had intended to found a monastery but his friend Bishop Oldham advised him to the contrary and the result was the founding of Corpus Christi College, Oxford. Both manor and mill were bestowed in 1529 upon the college who owned it until recent times, but Chertsey Abbey kept the tithe due upon it, until the Dissolution. The mill was of eighteenth-century date, with the adjoining house much older, while in 1929 an overshot wheel could be seen but no corn had been ground there for several years. Since that time the change has been considerable. A formal garden has been made around the waterways and as for the water-mill and house, these have been, externally at least, altered out of recognition.

Thorpe Mill, standing by a bridge below St Anne's Hill, after closing down in the early thirties, conserved its shell by becoming a dwelling and the side wing a tea room. Most of the weatherboarding has undergone repairs over the years. The general shape is sixteenth or seventeenth century. The undershot water-wheel is derelict and only the wooden spokes are left, all floats having been broken away. A small fall of water, only three feet, turned this wheel which represented the simplest construction and also exhibited the basic external position for placing a water-wheel. Little changed today after nearly forty years, though it was to become a well-known restaurant before its recent removal.

RIVER WEY

This river has the geographical inconvenience of having two branches bearing the same name, upon one of which stands Farnham. This town had six manorial mills, all contributing a substantial income to the holder in chief — the Bishop of Winchester, and later this revenue benefited from the growth of the corn market at Farnham in the eighteenth century. The first mill occurred on the west side of the town, called sometimes Willey Mill, then another nearer the town, while just south of the church a Georgian mill-house and mill-race survive of another. Hatch or East Mills are not old, and were converted into a laundry.

The Wey next drove High Mill but there are, before visiting this, two others, only a few yards from the river but supplied by ponds filled by springs and small brooks. The first, Bourne Mill, one of the largest for its date, sixteenth century, ceased working around 1928

Thorpe Mill, Egham (1933)

and later had become a hotel. Today it is a club, with the interior quite changed though the massive timber framing can still be seen. Unfortunately the two water-wheels in tandem were not kept, nor a magnificent clustered chimney stack at the back which has likewise disappeared. Though only a shell, all the plant having been removed when made into a hotel, it would have been the finest for its date as well as the largest. Its growth has been described. Much road alteration hereabouts, both new roads and road widening, has spoiled the surroundings.

The other watermill lies on the edge of the hill just above High Mill, or rather its ruined foundations in stone, brick and flint, also the position of the mill-race is recognizable, indicating sufficient height for at least a twenty-five foot diameter wheel. On the hill top can

Bourne Mill, Farnham, elevation to road (1930)

Side elevation showing water-wheels in tandem (1930)

Town mill, Godalming (1931)

Eashing Mill, Shackleford (1931)

be traced the wide lade for about a quarter mile, about forty feet or more above the valley, quite a stupendous earthwork feat especially if it represents the Medmille so frequently mentioned in the medieval periods.[8] The brickwork seems no older than the eighteenth century yet Rocque[9] does not show it nor does an early Ordnance Survey Map.[10] The Wey valley displays few sharp contours and a watermill perched up at this level must have presented an exceptional picture.

High Mill, named from 'le Heghe', lies in an unspoiled situation and with much of its machinery surviving; this makes it one of the best to be found in Surrey. Mill and house join under one roof, the former displaying internally quite an involved evolution, with the usual eighteenth-century brick lower storey with the floor binders and roof structures. As the shape of the latter indicates, these are of the same century's making, but some of the oak posts were retained from a previous mill, probably fifteenth or sixteenth century, and the pine rafters rise off older oak plates. The plant constitutes one of the delights of the building. In addition to the present water-wheel there are suggestions of a second water-wheel on the south side, following a common practice in this region, which may have succumbed in an early nineteenth-century alteration. The clasp-arm pit-wheel is a choice wooden specimen driven by a cast iron water-wheel with wooden plashers. The rest of the gear wheels are also wood, driving two sets of stones, of which only one lower bedstone has been kept.

The Wey next supplied the power for the mill attached to Waverley Abbey, a Cistercian foundation whose ruins can be seen from the bridge where the mill stood until 1900. Today, weirs and arches in flintwork and clunch denote the remains of a large structure, which originally may have been put there for the use of the monastery. Next, at Tilford Bridge, a cottage and weir mark another vanished mill. The Wey joining up with its twin branch, flows to Elstead Mill, a tall object in this countryside. This example, probably representing the site of one of the six credited to Farnham, may also be the site of the nineteenth-century paper-mill once here, which would have later housed a 'fringe making' factory and later still produced flour. The present building cannot be that one which replaced the mill burned down here in 1649.[11] It has been a dwelling house for a long time.

On a small tributary and pond stood Cutt Mill, Puttenham, removed in the thirties, while Pepperharrow Mill, a Domesday foundation, must have stood at Somerset Bridge where a weir exists still, probably never rebuilt from its ruinous state as recorded in 1353. The Wey next drove Eashing Mill which, apart from its situation, had little charm, or age. A predecessor with four water-wheels ground corn until 1658 when it was converted into a paper-mill. William West of Wraysbury, Buckinghamshire (where there were also paper-mills), owned Eashing Mill during the same century. However, by 1704, corn-grinding plant had been installed alongside the paper-making plant and both remained until 1850, the date of most of the present buildings.

On the way to the mill below Godalming station, the river passes the site of a vanished mill once belonging to Hurtmore Manor. After this a Victorian example, Salgasson Flour Mill, and the Leather Mill could, until a few years ago, be seen from the railway. However, a new factory has taken over both sites, the Leather Mill occupying the original position and Salgasson Mill probably a fresh one. The Wey flows to Catteshall Mill, where a castellated Victorian mill marks a site, more with curiosity than elegance. All the same it is of Domesday origin, and was bestowed by the holder Geoffrey Purcell, the King's Usher, upon the Abbey of Reading when he became a monk there. Centuries later cloth-fulling was added to its activities but in 1699 the mill was converted to paper-making and produced the 'white-browne' paper, a noted Godalming product, used chiefly by grocers. Another, Unstead Mill, also a Domesday site though the original position may have been altered when this part of the Wey Canal was formed in 1760, has been for more than a century like a factory, though an earlier one may have been that figuring in a quarrel between Robert de Chisenhall and the men of Shalford in 1379.

The Wey received a tributary from Loseley Park, where Polstead Mill could have been, though no sign of it exists today; probably the landscaping and emparking done by Sir Christopher More in 1535 may have caused its removal. The Wey now approaches Guildford, but the third mill attached to Shalford, owned by a member of the De La Poyle family, adjoined the river alongside a bridge, which was of much convenience to travellers visiting Shalford Fair, who wished to pay their devotions at the chapel on this hill. Next comes Guildford. In the early thirties the vicinity of its great mill and waterways below St Mary's Church, together with the roofscape as one looked up towards the town, presented as worthy a picture as did the High Street. Today this prospect has become a much less engaging sight. Large departmental stores, a new theatre, and a new road at river level, have been merciless towards the former subtleties of this picture.

The great mill built in 1766 keeps its identity, externally at least. The newly-formed entrance to the Yvonne Arnaud Theatre has opened up the approach to the building. Part was rebuilt and probably extended in 1896 in connection with the waterworks, a use of long standing. When this was carried out the same shape was continued with steel girders replacing older timber members, though inside much of the original pine framing can be seen. The part not used for the pumping machinery seems to be used as a store room and workshop for the theatre. Records frequently mention the town's watermills, though identification between mention and particular site can be confusing or obscure.

Apparently on the western branch of the Wey near

Millmead stood Artington Mill but it vanished centuries ago. It was attached to Artington Manor, originally part of Godalming, and in the early days of Henry III's reign belonged to Stephen de Turnham, whose daughter, Beatrice de Fay, afterwards retained a fourth portion. Mills on the site of the present example were evidently the soke mills belonging to the manor of Poyle (the name survives in Pewley Hill half a mile away) which William the Conqueror presented to Robert Testard. In the thirteenth century Geoffrey de Braboeuf and Richard Testard, obviously a descendant of the above Robert, owned certain (probably two) of the mills, and Richard also possessed the two castle mills adjoining to the north which were later sold to the Crown.

Before 1256-7 some incompetence, or ill advice, caused the Crown to disuse the castle mills, and build a corn-, a fulling- and a malt-mill in a fresh position below Guildford bridge, near the park. The new structure brought about the flooding of the lands of the joint owners of the manor of Artington, besides impeding the working of Artington Mill. Obviously the mills stood too close to each other and a back flow from one pond tailed the wheel of the other. The king decided to respect popular opinion and immediately sold the newly-erected mills to Geoffrey de Braboeuf and Richard Testard, who presumably removed them. Evidently a set of watermills with two wheels for corn grinding and two for fulling were built afterwards, for that number belonged to Poyle Manor in 1439, with which they remained as late as 1643 when Henry Smith gave from them £40 a year for the poor of Guildford. In 1649 they still ground corn and fulled cloth, and they are depicted in an engraving of 1690 from the Royal Magazine.[12] Four wheels are shown: two large wheels were attached on the northerly mills, evidently the corn ones, and two smaller wheels, one on either side of a lower building, fulled the cloth. Probably one of these was that set aside in 1701 to pump water for supplying the town, when a piece of land was also reserved for the 'engine', doubtless the storage tanks. According to Buck's view of Guildford (*c.* 1738) the group had suffered no change and probably remained intact until the re-building of 1766. On that occasion two wheels were kept for the grinding of corn, a third for fulling and a fourth for the waterworks, but by 1800 the fulling apparatus was superseded by corn grinding.

The Wey's next mill, that at Stoke, north of Guildford, is Victorian and something of an eyesore, or so it seemed in the thirties (the enthusiasms of today not then being envisaged). It stands a short distance from the road, a position which may be explained by its history. Now sited south of the Wey, the original building stood on the north side and the road, as a ford, passed over the tail. However, in the twelfth century the Bishop of London, the landlord, demolished the mill and rebuilt it on the south bank of the Wey, and this time to the east of the ford with the consequences that would be expected. The dam alongside the mill held up the water, and in a short space of time the ford became so deep as to be too dangerous to use. The neighbouring folk grew angry and contemplated overturning the new mill, but the bishop, wishing to allay any such popular demonstrations, built a bridge at the ford at his own expense, aided by the alms of other people, and thus eliminated the menace. The bridge, or a successor, evidently made of wood, survived until 1780, when Stoke Park and mill were purchased by a Mr Aldersey who diverted the road from the original route past the east of the church and over the ford, to its present position. Later maps show a paper-mill south of Stoke corn-mill, for which former, the old position was re-utilized.

Bowers Mill, Burpham, overlooks the lock on the canal. Though the site dates from ancient times, a new mill, probably on a new site, was built about 1651 when the canal connecting the Thames and Guildford, an idea planned by Sir Richard Weston a decade earlier, had been partly completed. The existing cottage may belong to this mill because the latter was erected only last century.

Apparently near Triggs Lock at Sutton Green stood a fulling-mill, attached in the fourteenth century to the Royal Palace at Woking. As Aubrey's map of Surrey does not show it, it must have vanished earlier. Near Old Woking can be found the St. Martha's Printing Works, the site of a watermill belonging to the royal house here. A façade with eleven gables, the only portion worth a glance, still looks into the pond, a picture much as in the late twenties though the tail pool seems to have shrunk in extent. Hereabouts the Wey is joined by a rivulet coming from Clandon, and in the park of that name there stood until the eighteenth century, a watermill. Today only the outline of the dam can be recognized.

The next on the Wey near Newark Priory, and of the same name, provided the finest example, pictorially, in the Home Counties even though it had lost much of its plant.[13] Following its closure, *c.* 1952, great hopes were entertained of saving the building and at least one of the plants in a conversion to a restaurant, but all such deliberations were cut short in December 1966, when fire left nothing but two water-wheels, some brick walls and a heap of charred wood. An earlier mill here was given by Thomas and Alice de Send to Newark Priory and after the Dissolution, it probably went to Sir Anthony Browne. This one probably existed until about 1650 when the formation of the canal brought about its destruction. The canal makers would have utilized the existing lade and at this time the present Newark Mill would have been built.

From this site the Wey flows in about three arms, and one of these, with the aid of a small tributary from Effingham, filled the pond for Ockham Mill, working until after the last war and apparently still retaining part of its plant. The date, 1862, when the local squire built it, partly explains why by size and design it should be so unlike any contemporary country mill,

High Mill, Farnham (1930)

Bowers Mill, Burpham, Surrey (1931)

Newark Mill, Ripley, Surrey (1931)

but of its designer there seems no record. Also its appearance, rather bizarre to see across Surrey meadow-land, and more like a transplant from old Lübeck, has weathered well and would seem more acceptable now than it did in the twenties. Ockham Court, the manor house, lies across the pool, a closeness that matches their long association in history from the time when it was a double mill, worth £2, and held by Gilbert de Clare. The river would next be driving Wisley Mill if this latter had not long vanished. The name of Mill Lands near Wisley Bridge suggests a site.

The Wey leads to Byfleet where the mill of that manor, or its last representative, of eighteenth-century date, stood south of the Byfleet–Cobham Road. The Domesday mill here appears to have been royal property, but in the thirteenth century it was a double mill and belonged to Chertsey Abbey. In 1284 the mill had been demolished. The tenant Geoffrey de Lucas paid 18*s*. for the site, and for the fishery 3*s*., while the miller's rent stood at a nominal one shilling. Evelyn has in his diary for 24 August 1678 the following entry: 'went to see my lord of St Albans House at Byfleet, an old large building. Thence to papermills where I found there making a coarse white paper . . .'. As Aubrey's map of Surrey shows an iron-mill at the head of two large ponds at Byfleet, the paper-mills could not have survived for long afterwards. In the early 19th century it became a corn mill. Much of this structure survives, but only as a store though well looked after. The plant was removed during the last war, leaving only the upright shaft and part of the hurst that belonged to the first water-wheel. The second water-wheel on the east side retains only the cast iron axle and the hurst columns of the same material. With the fine 19th century mansion of the miller it forms an attractive group, well supported visually by trees and waterway.

A modern flour factory, Cox's Lock Mill, by the canal (the latter probably following an earlier channel) downstream of the railway bridge, represents the site of oneof two mills given by Henry III in 1235 to William, son of Daniel Pincerna, one of the unpopular

Ockham Mill, Surrey (1931)

'foreigners' of that reign. This mill was 'above' bridge, while a second stood at 'Freyreford', the site likely enough of that shown to be[14] there in the mid-eighteenth century, where the Wey joins the Thames, later represented by Ham Mills, where only some pieces of Victorian brickwork survive.

NOTES AND REFERENCES

1 Hindley and Crostly, *Map of the County of Surrey, 1789-90*, London, 1790
2 For picture see Hopkins, Thurston, *Old Watermills and Windmills*, P. Allan and Co., London, 1930
3 Rocque, *Map of Surrey*, 1768
4 For description of interior when working see, Hillier, op. cit.
5 A phrase he may have borrowed from Aubrey, J., *Perambulations in Surrey*, E. Curll, London, 1719
6 Aubrey, op. cit.
7 Hardy, W. J., (ed.), *Home Counties Magazine*, vol. 5
8 Robo, op. cit.
9 Rocque, *Map of Surrey*
10 *Surrey*, ed. of 1807
11 See *Surrey Archaeological Collections* VII. 193
12 Williamson, G. C., *Guildford in Olden Time*, Bell, 1902
13 Hillier, op. cit., gives a description before its removal. See also Vol. I, pp. 94-103
14 Rocque, *Map of Middlesex*, 1754

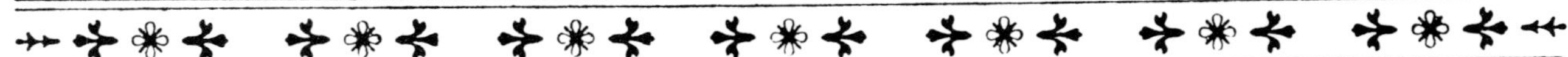

Chapter 13

THE WYE, BRENT AND CRANE

RIVER WYE

THIS river is second only to the Kentish Loose in the league of mills per mile of river, Wandle coming a close third. This record stands from medieval times, though by last century some of the northern streams might have surpassed both. It is not easy to identify precisely a mention in manorial and town records with a particular site on the Wye, for at least twenty-five existed before it joined the Thames.

The first, at the edge of West Wycombe Park, a curious brick and flint structure that housed a circular saw, seems to have been removed. Probably built when the Dashwoods had their park landscaped in the eighteenth century, the site did not look to be a really old one. The Wye next worked Little Mill which has long since ceased working and though the foundations of the mill, the miller's house and pool remained in the early thirties, these appear now to have gone.

One loss was that of Lower or Fryers Mill that stood near the present recreation ground. A late sixteenth- or seventeenth-century building with silver-hued weather-boarding, and a brown tiled roof next to a fine Georgian miller's cottage, it formed a feature that surely should have been retained, with the attractive ford across the Wye. It doubtless represented the mill where Thomas Parnell, 'miller of Brook', was fined by the manorial court in 1505 for profiteering. He still worked there in 1540.

The last on the stream in West Wycombe could be found just below the corner at Oakridge Road. At first glance the building looks anything but a watermill; an asbestos roof and wooden-windowed front hardly suggesting the origin. Since the last war, demolitions permit one to see more easily to the rear, where a wooden water-wheel axle has been built into the concrete walls now enclosing the wheel-race. All the same, the structure dates only to mid- or early-nineteenth century. This example, latterly called Lords Mill, had borne the name of Paper Mill in the seventeenth century, Balls Mill in 1796, and Nownes Mill in 1825.

The Wye now comes to the many mills of High Wycombe and it may be apposite at this point to speak of the town which long ago pulled ahead of Buckingham and Aylesbury, while the Slough conurbation is but an upstart. Tradition cites the river Wye as the cause. This stream provided plentiful and convenient power for an imposing array of watermills. Their presence drew to the town trade above the normal activity of an agricultural market. Here was the base upon which grew High Wycombe's prosperity.

Domesday records some six watermills, valued at seventy-five shillings, and in addition the manor of Lude, then without the town boundaries, possessed another three. Merchants settled in the town during the early twelfth century while, in the civil wars of Stephen's reign, the mills were damaged, probably in an assault on the castle which now no longer exists. Nevertheless the Crown granted the holder £4 indemnity, 'In defecte molendinorum' (an early war damage claim!), and the millers continued their craft, producing the flour used for the fine white bread described in those days as a Wycombe speciality.

Some of the present mills were established before Domesday and the remainder by the early thirteenth century, for at that time the abbey of Godstow, in holding the rectory manor and advowson, claimed tithes from some nine mills. The cloth-making industry flourished in High Wycombe during this period, when commences the mention of fulling-mills. Prosperity must have been considerable, as the town did not, to any great extent, increase its size until the late eighteenth century.

Cloth as a manufacture seems to have died out by the sixteenth century, but its place was partly taken by paper-making. This trade in 1690 kept some eight paper-mills going and supported some fifty families of workers. Today not one works at its original job, and most have either been pulled down or are hidden by

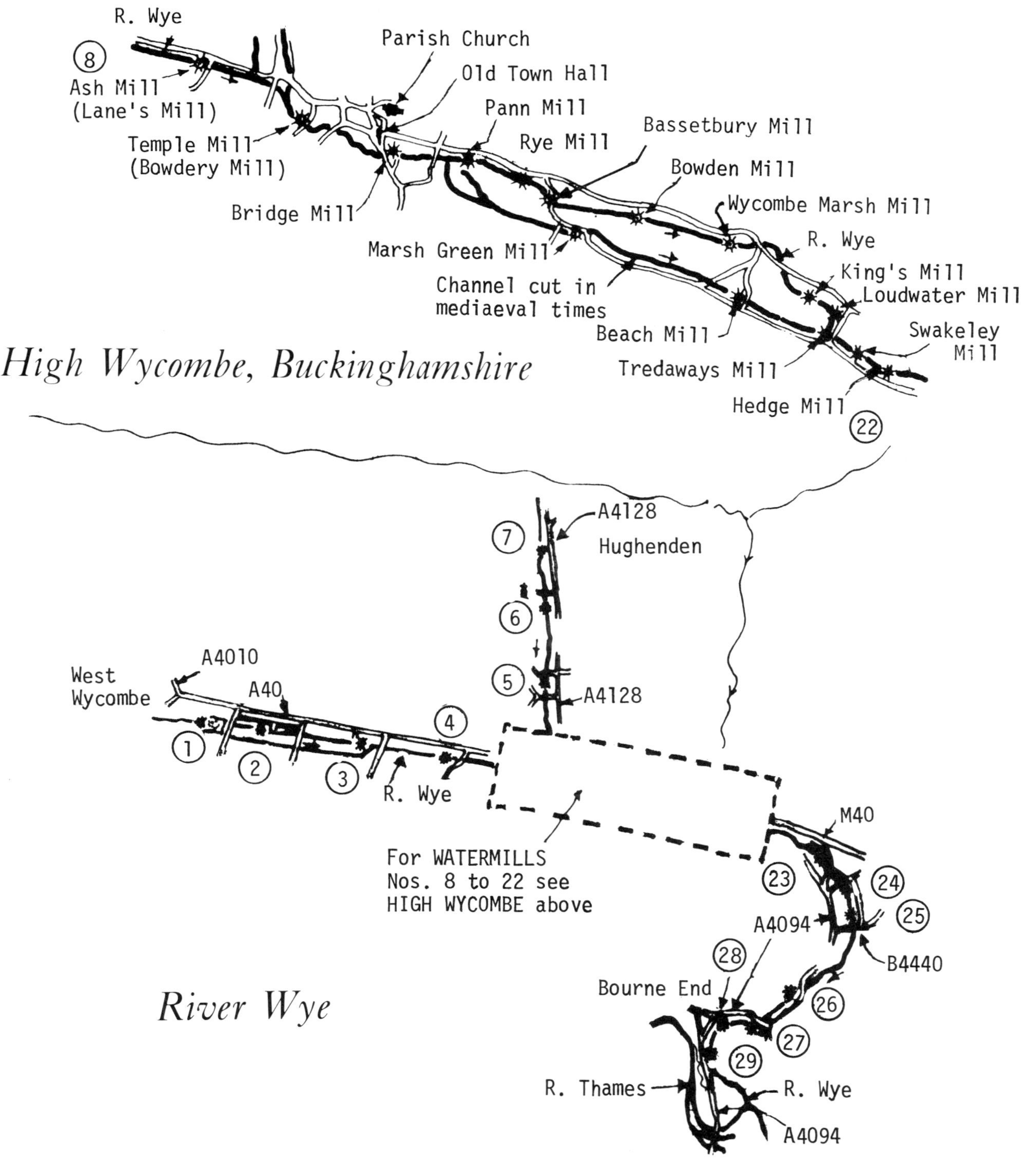

1 West Wycombe Mill
2 Little Mill
3 Fryers Mill
4 Lords Mill
5 mill site
6 Flint Mill
7 mill site
8/22 as in map above

23 Clopton Mill
24 Old Glory Mill
25 Lower Glory Mill
26 Soho Mill
27 Princes Mill
28 Bourne Mill
29 Hedsor Mill

later factories, converted into dwellings or into stores, or left derelict.

As late as the fifties some dilapidated lower walls were the only remains of the first mill which stood at the corner of Westbourne Street and the Oxford Road. Before its demolition it was mainly an eighteenth-century construction with a weather-boarded upper storey upon brick. It had a long recorded pedigree, representing no doubt one of the Domesday mills. Today the site is a carpet of rubble, while the mill-race survives represented by a concrete spillway, over which the Wye (here in an artificial channel) flows unexpectedly clean, considering how garbage-laden is the lade leading to it. An early mill on this site was given in 1200 by Mabel, daughter of Siward, to Godstow Nunnery. Later it can be seen taking various names, probably of owners: Broughton Mill in medieval times; Wynkles Mill in the sixteenth century; Lanes Mill in 1796; and Ash Mill in 1878. Until not long ago the tail formed a feature as a stream following the line of the Oxford Road.

Below the last mill the Wye receives a tributary from Hughenden. The latter drove a long vanished water-mill somewhere in the park south of the church, or maybe the Flint Mill shown on a map of 1832. The Wye flows to the next, Gosenham or Temple Mills, whose site, recognizable only with difficulty in 1952, could once be found down Bowdery Lane. Of this part of High Wycombe the less said the better, vestiges of antiquity, as well as other things, having been obliterated. The name derives from the Knights Templars, who received a portion of the important manor of Wycombe from Roger Vipont in 1227. These mills were used for fulling cloth in the fifteenth century, and remaining in the same trade, were in the hands of the Rance family in 1623. The mill was taken down long ago though the site of the wheel-race could be made out in the thirties.

The Wye then went its way, over a heterogeneous collection of rubbish to a fine old watermill standing until 1932, in St Mary's Street. Then a disastrous fire made short work of its sixteenth-century beams and trusses.

> After the fire, what an appearance greets the eye now! A mass of fire-blackened ruins, roofs fallen in, immense oak beams charred five inches deep, millstones perched perilously on half consumed timbers, the iron water-wheel buried in debris and the whole place deserted, while even the large trees, which shaded the tail, have been badly singed. When a flour mill catches alight, everything inside feeds the flames and the heat generated within a few minutes is so great that the normal easy access to plenty of water becomes virtually useless.[1]

Bridge Mill, as this one was named, formed part of the endowment of the collegiate church in Wallingford Castle (a building said to have been, until its destruction in 1649, a rival in size and beauty to St George's, Windsor). A car park and the legion hall now occupy the site, while the lade, though dry, can still be identified.

The next mill on the main branch of the Wye is Pann or Pannell's Mill, a name coming from the twelfth-century family of Pinel. The present mill ground corn until quite recently. Its three gables would represent an early seventeenth-century building, when it was in the ownership of Christopher Wase.

With a recollection of its former appearance one first notes that some post-war reparation has replaced the former white painted weather-boarding with fletton brickwork, a misfortune that has taken away so much of the building's former character. The owners must have rebuilt the interior as well some time after the last war, and strangely enough, retained the rafters of the old roof and some smaller portions of half timbering, so that the original shape can be observed. That is about all. The plant has now been removed. A large cast iron framed water-wheel at the rear, drove an iron pit-wheel which then turned a line shaft, and some of the pulleys that powered the reducing machines remain. The present wheel represents a resiting from the preceding one, which before its removal, stood in the centre of the mill building.

Rye Mill, along the London Road, follows next in sequence. It once belonged to Temple Wycombe manor, but as it is an industrial building of little beauty, we pass on further down to Bassetsbury Mill. This one lies in close proximity to a fine manor house, a juxtaposition which aptly expresses their close association in history. It ceased its original occupation long ago, but in 1932 the owner converted it into a showroom for antiques. By then much of the plant had been removed to accord with its new use, for in those days things like cogs, gear wheels, and flour-dressing machinery did not come within the meaning of antiques. The water-wheel, which old deeds indicate stood amid the mill, had disappeared long before 1932.

The building itself is the work of several ages, an evolution of addition and alteration from a small sixteenth-century mill, much like the cottage now at the eastern end. The first alteration was carried out, probably by the Loggan family, *c.* 1679-91, when the half timber walling was encased in brick. Later alterations included new windows and work to the roof. More recently another set of alterations has taken place. Fortunately, by this conversion to several dwellings, the Bassetsbury Mill survives as a pleasant enough domestic group, though rather a husk, denuded of its original significance. By an imaginative gesture that great cast iron water-wheel stands at the rear, a feature in itself and a much appreciated link.

Lower down lies Bowden Mill. No doubt its name comes from a Wycombe lady who left considerable sums to local charity. The original portion with one water-wheel would date possibly from the sixteenth century, while much later a wing was reconstructed, together with another wheel, echelon-wise to the original. An early fulling-mill belonged to a certain

Fryers Mill, West Wycombe (1932)

Pann Mill, High Wycombe (1932)

Bowden Mill, High Wycombe (1932)

Henry Cheesemonger in 1235 and contained two water-wheels and probably occupied the site of Bowden Mill. As late as 1950 new building had not encroached sufficiently to spoil the scene. Hardly so today — the miller's house has gone, suburbia has crept up on one side and the sewage works on the other. The best feature lies in the still pellucid Wye, which falls over the race of the removed water-wheel. A remaining wheel, cast iron on a wooden axle, still appeals, which cannot be said of what remains of the building itself, today merely a store for a garage.

A mill at Wycombe Marsh was called, in the twelfth

century, 'Gwynaunts Mulne' and was bestowed upon Missenden Abbey by Ellis Gwynaunt. No remains survive of the pre-factory mill except two Georgian houses of striking appearance. Next comes King's Mill where the late eighteenth-century mill building and surroundings have been industrialized.

Loudwater Mill, a vast paper factory today, had an early predecessor recorded in Domesday. Subsequently it was rebuilt as a fulling-mill by Thomas de la Lude (hence the name), for during 1250-60 Philip Basset, lord of Bassetsbury Manor, granted him permission to do so at a rent of one mark yearly. One condition was attached, namely, that no cloth made in the borough of Wycombe was to be fulled there. Philip obviously wished to reserve that lucrative trade for his own fulling-mill, Bowden Mill. Loudwater Mill, described as possessing two water-wheels, a common plan in this valley, remained with the De la Lude family until 1337 and by the Dissolution had come into the hands of the Knights Hospitallers.

Before the Wye left Wycombe it drove two more — Swakeleys and Hedge Mills, paper-mills since the seventeenth century. They have lost their identities in the vast factory which covers their sites, and produces Ford's blotting-paper.

A southern branch of the Wye, called for a part of its length the Dyke, or New Water, leaves the main course below Bridge Mill, and flows for nearly three miles along a minor engineering feat, constructed, I would suggest, in the early seventeenth century, to supply the new paper-mills. This may be too ready an assumption, as the work could date from earlier. It leads first to Marsh Green Mill which in part might date from about that time. Once with a large pool (now filled in) it lay in 'Corotesque' surroundings set off by waterways and the usual farm accompaniment of orchards, gardens and outbuildings. The roar of well churned-up water issuing from the water-wheel, so often audible when this example was visited in the early thirties, had ceased after the last war and the surroundings had declined in attractiveness. The group, the miller's house, and farm building, have been transformed into dwellings, quite pleasantly too, but the mill itself seems not to have survived, the position being taken by the garages.

Further down, a pleasant ford can, or could be, found, marking the site of Beech Mill, taken down around 1900 for drainage works. The next and last example, standing near the Loudwater Chapel, was Tredaways or Overshot Mill, which as late as the 1930s retained both general shape and details from its seventeenth-century origin. Only the wheel-race recalls this one.

Clapton Mill, Wooburn Moor, under the brow of a new motorway viaduct retains a nineteenth-century structure, obviously not surviving from the wire-drawing mill of 1796 but probably part of a tarpaulin factory here *c.* 1875. Then comes Old Glory Mill, named from a thirteenth-century holder, Robert Glorie, and further down the river, Lower Glory Mill. As late as 1930, some eighteenth-century walling remained of each, but this is not so today. Both these, as paper-mills, were leased at a peppercorn rent to the seventeenth-century poet, Edmund Waller. The next, Soho Mill, represented by a factory, dominates what remains of Wooburn village. Princes Mill in Hedsor, the next parish, and likewise an old site, has gone, while Bourne Mill was a malt-mill in 1555. Nearer modern times, it made explosives and subsequently mill boards, but it has also been removed. This leaves Hedsor Mill, the last on the Wye, and which still produced flour in the thirties, using two undershot water-wheels with iron frames and wooden floats. It has given way to a new bridge.

RIVER BRENT

Hendon has Mill Hill which proclaims its windmill, but a watermill also existed. Rocque[2] gives us no information on his map and the site remains uncertain. That he does not show it suggests that it had disappeared by the eighteenth century. Either a waterfall near Brent bridge, or another at Renters Farm near the sewage works, may mark possible sites and the probability lies with the latter, as this is near Clitterhouse farm. Records state that Goldherds Mill and, in the early fifteenth century, Goddards Mill (probably the family of Goders who owned it and from whose name Golders Green descends) was described as near Clitherhouse.

On the other hand records also mention a mill at Edgware Bridge. Domesday records one at Kingsbury. As the boundary of each manor coincided along Watling Street the possibility arises that these two references represent the same mill; also that it might have even been the site of Golders Mill must be considered. However, old maps do not help while the formation of the Welsh Harp destroyed possible clues.

Hanwell had a watermill in addition to the manorial windmill. The former remained during the last century at Dormans Well on Hanwell Common, presenting a pleasing picture with a pool and the accompanying trees. An old illustration[3] depicts it as a piece of homely seventeenth-century building. A tributary of the Brent runs through Osterley Park which was owned by Sir Thomas Gresham in the late sixteenth century. Within his grounds he built a water corn-mill, seemingly in 1571, though likely enough the previous manorial lord, the Abbess of Syon, held one there previously. Before his death in 1579, he 'ioned a paper myll theronto and yet vsed the same myll a corne myll still, and all vnderone roufe and dryven by one Steame'. Norden, in an account of the Park and mansion says, 'a faire and stately building of bricks . . . garnished with manie faire ponds which afforded not only fish . . . but also great use for milles; as paper milles, oyle-milles and corne mill, all which are now decaied (a corne mill excepted)'.

Hedsor, Buckinghamshire (1933)

Isleworth Mill, Middlesex (1931)

Marsh Green Mill, High Wycombe (1932)

An inquiry was set forth in 1584 to decide whether Gresham's mill had encroached on to the King's highway, when its situation was described as upon the Brent near 'Cruxellesforde'. This could not be correct because the mill depended upon a small stream which rose in the park, filled the ponds mentioned by Norden, and then, after turning the wheels, joined the Brent. By 1754 only the pool remained.[4]

At Brentford no mills were to be found in Domesday but latterly two are recorded. One which stood on the Thames bank near Smiths Hill was probably the predecessor of that one conspicuous for its height, and to be seen early last century. The second occupied the site of the Canal Boatman's hall at the bottom of the Butts adjoining the weir, which latter, and presumably the mill as well, belonged as early as 1313 to the canons of St Pauls. However the weir was moved in response to a petition of 1750 but the mill, a plain gabled and tiled structure of the seventeenth century, remained for over a hundred years afterwards.[5]

RIVER CRANE

The Yeading Brook rises south of Harrow, later becomes the Crane and joins the Thames at Isleworth. During the middle ages some land owner, said to be a fifteenth-century abbess of Syon, but more likely an earlier predecessor, tapped the Colne to obtain an additional supply and conveyed the water to the mills at Isleworth. Apparently another channel became necessary later and records refer to the making of this about 1546. This was the Duke of Northumberland's river which for much of its way utilized existing channels. The additional volume of water furnished to the Crane made possible more mill sites, no doubt one of the motives as well being to provide a supply to Hampton Court. The first two of the sites lay upstream of Baber Bridge at Hounslow Heath. One of these, probably that nearer the bridge, may derive from one credited in 1375 to the manor of Imbury whose channel fed the sword-mill mentioned in 1635. It seems to have been used for paper-making in the early seventeenth century by a certain William Bushes, who was indicted before the county sessions in 1636, 'for grinding ragges' at his mill. One of his servants contracted the plague from old clothes and suchlike that the latter collected in London and the consequent, popularly-expressed, alarm forced him to shut down his mill. This establishment carried on because an early eighteenth-century map shows it, by which time a powder-mill had arisen upstream and no doubt took over the paper-mill site.

William Mason, poet and friend of Horace Walpole, mentions these gunpowder-mills in his 'Heroic Epistle', a satire upon the architect, Sir William Chambers.

> Hounslow, whose Heath sublimer terror fills,
> shall with her gibbets, lend her powder mills . . .

Among the explosions frequently recorded, as commonly happened in these places, one in 1772 damaged part of Walpole's house at Twickenham, his 'Gothic' Strawberry Hill three and a half miles away, while a more expansive contemporary claimed that the bang (probably from the gunpowder-mills situated further downstream) was also heard at Gloucester. The Government closed down the works in 1926, and by 1932 little of the structures survived, and, in the rather dreary prospect then presented, the pond, spillway, and a Gorgian house, summed up all that was visible from the road. Today one can easily tresspass over the site, a dirty, dreary piece of woodland with a few walls and channels to indicate the past, and can observe one wheel-race with an opening alongside in the brickwork where the axle turned.

A 'swordsmill' which later became a 'Brazil' mill stood on the other (south) side of the Roman road by Baber Bridge and survived for over a century, though now no more. Other gunpowder-mills could be found a couple of miles downstream, the site being shown on an eighteenth-century map.[6] In 1869 a heavy explosion occurred, and later it seems they were used for paper-making. This site seems to be of fourteenth-century origin; occupied by a manorial mill for Twickenham, being replaced later, in 1768, by the gunpowder-mills which in turn no longer exist. The Crane was dammed up further down its course in the eighteenth century to drive two mills at Feltwell, and after the river divides into its two branches, two other examples were established. These were calico-mills, built in 1769. The last one, formerly standing alongside the Royal Oak in Worton Road, has left no indication of its existence; the water courses have been altered subsequently. The building was a fair-sized one with a large water-wheel.[7]

One of the branches flows on to drive the mills once as Isleworth,[8] and the other direct to the Thames. On the former, near St John's Church where the river takes a bend, stood a mill from early days. After the customary use for corn-grinding, it succumbed to the lure of the London market, in that a partnership of four men established a brass-working establishment on the site in 1582. Its original is said to have been built in 1534, but no valid reason exists why an earlier predecessor could not have been one of those belonging to Isleworth manor. It seems to have been noteworthy enough to have impressed John Norden who says 'a copper and brasse mill it is wrought out of the oar melted and forged . . . where workmen make plates both of copper and brasse of all syces, little and great, thick and thyn for all purposes, they make also kyttles.'.

Equipped with tilt-hammers, it would have resembled the Sussex iron-mills, for the same authority adds 'some hammers of wrought and beaten iron, some of cast iron of 200, 300 and some 400 [lb] weight . . . lifted up by an artificial engine by force of the water . . .' It remained a copper-mill in 1649 and was named 'Brassel Mill' in 1740, though by then probably

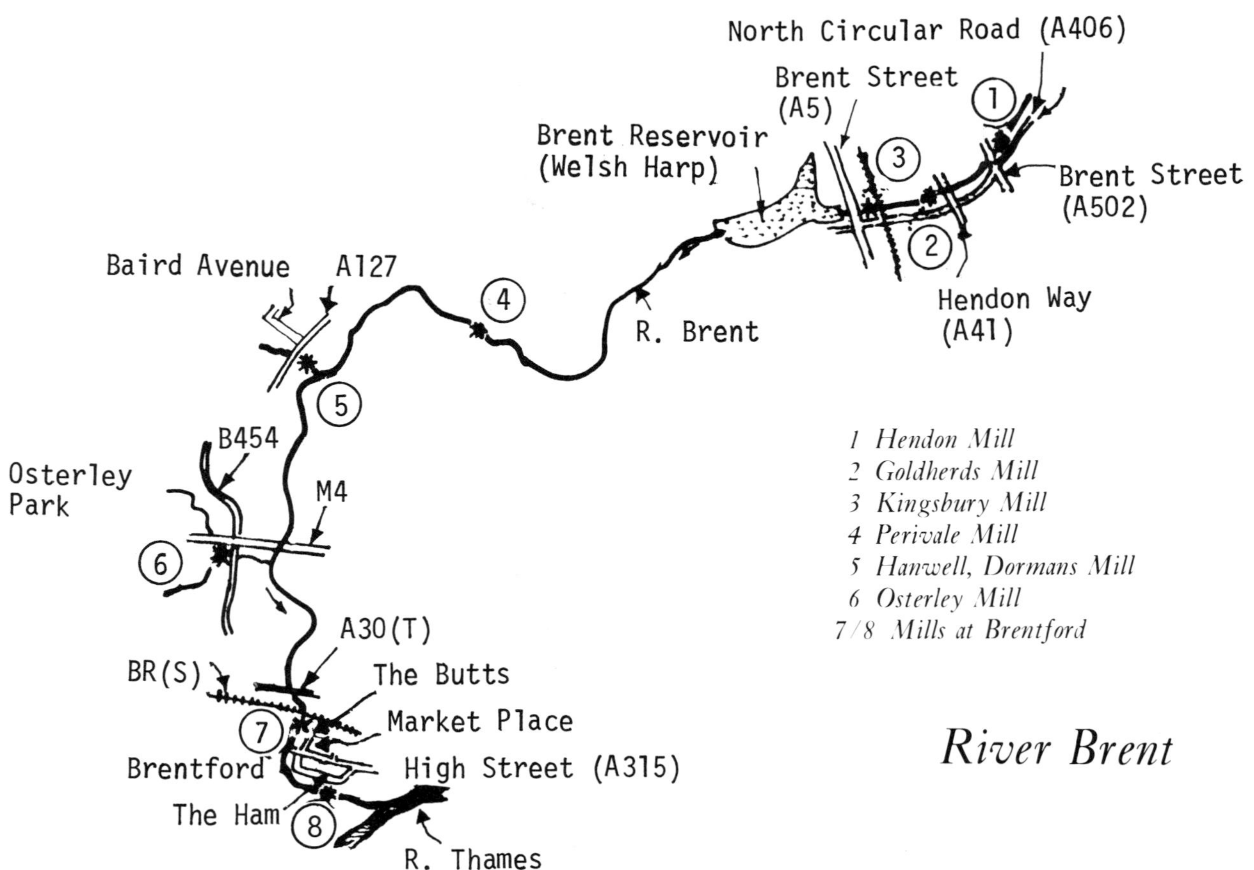

River Brent

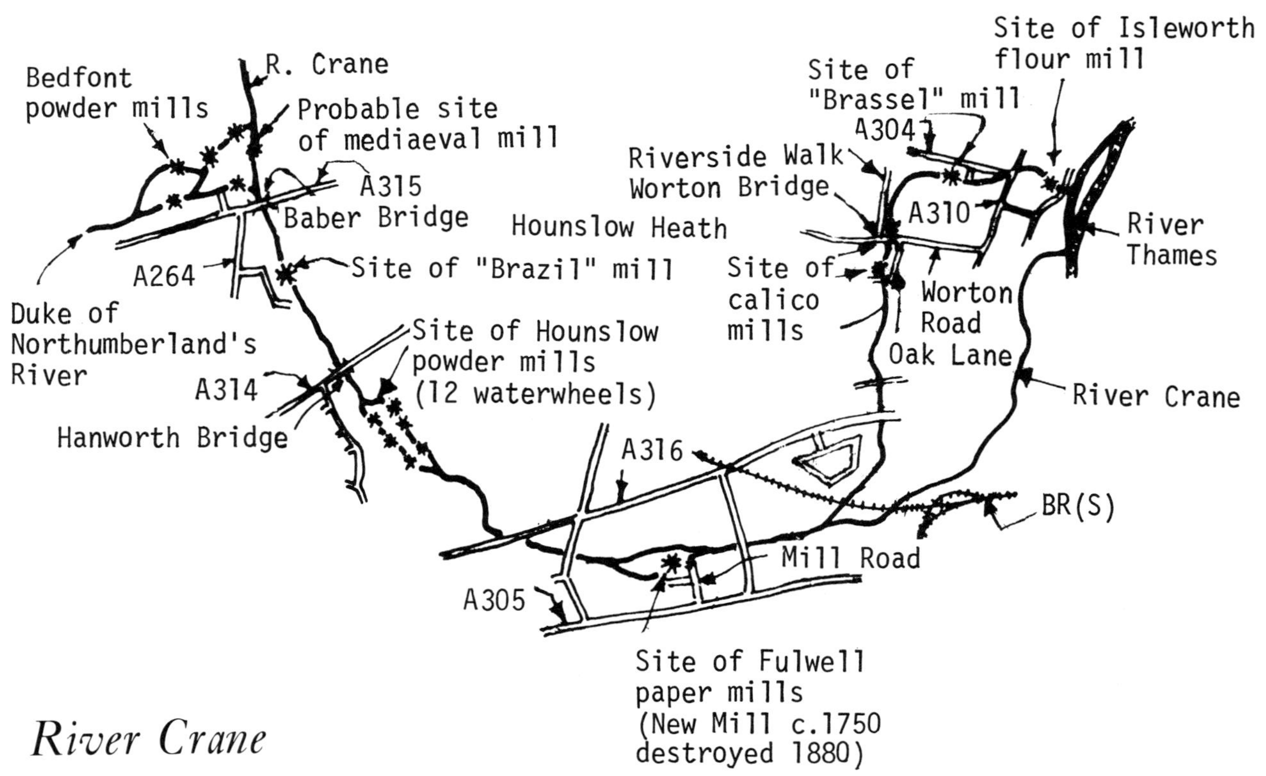

River Crane

out of use. Today, from the bridge at the entry to the Brewery Yard, one can see the weir, now much reduced in fall, while the pleasant eighteenth-century warehouse building facing on to St John's Road may be a survival from the manufactory. In fact a print of 1635 gives a birds'-eye view, if maybe a somewhat distant one, of this establishment, which consisted mainly of two long parallel buildings, and the present part could be a latter replacement.

Before 1939 the last mill on the Crane lay at the end of Church Street, a truly immense structural complex fronted by the early nineteenth-century watermill. Today it has vanished, and the blocks of flats, garages, and offices newly put up all around the site emphasize the fact of its removal. The tail came through the older building in three culverts, two of which have been built up. These emptied into a large pool contained within encircling brick walls and forming at high tide a basin for barges to approach right underneath the triple grain hoists through which the grain sacks were hauled up to bin floor level. On the site this process can still be visualized despite the befouled state of the water. A fine single arch carrying the roadway, and no doubt contemporary with the destroyed mill, should not be overlooked.

The last two mills represent those appurtenant to the manor of Isleworth, which in the thirteenth century belonged to Richard, king of the Romans. He being naturally of the king's party during that civil war which culminated in the battle of Evesham, his property offered an obvious target in 1263 when a London mob, led by Hugh le Despencer, marched to join Simon de Montfort and his followers at Isleworth Park. They 'spoiled' Richard's house on their way and destroyed his watermills. Later replacements became part of the endowment of Syon Abbey at its foundation.

NOTES AND REFERENCES

1 From the author's notes taken in 1932
2 Rocque, *Map of Middlesex*, 1754
3 Philimore, W. P. W., (ed.), *London and Middlesex Note-book*, E. Stork, London, 1892
4 Rocque, op. cit.
5 Turner, F. A., *History and Antiquities of Brentford*, W. Pearce and Co., Brentford, 1922
6 Morden's *Map of Twenty Miles around London*, 1730 (East of Hanworth Bridge)
7 For views of this and other mills on the Crane see B M Maps xxxi c. 5 g and xxxi c. 5 h, Isleworth (Brassel) Mill. For other Isleworth Mills see B M Maps xxx 19 ii and xxx 99 hh
8 For Isleworth Mill see Bate, G. E., *And so to make a city here*, Thomasons, Hounslow, 1941, p. 224, 278–82, 299–302

Chapter 14

THE OUSE AND TRIBUTARIES

WITHIN the area chosen some rivers, which eventually joined the Great Ouse, carried watermills and these are grouped into this short chapter. The first is the Ouzel and its head streams. On one of these, at Ivinghoe below the green, stood a watermill called at one time Pightlesthone Mill, and then part of the manor of Pitstone. With the other mill at Ford End, it was recorded in 1231 as being handed to one William, son of Herbert. Ivinghoe Mill was acquired by Henry Spigurnel, probably the noted judge, from John de la Bere, a tenant of Ralph of Ivinghoe — a transaction, as the latter held direct from the king, of an unlawful nature, which had to be rectified by a pardon in 1306. In 1447 it was called Ivinghoe Mill and then held by Anne Kirkham, heiress to the Spigurnel property. Town, Pitstone or Broadmead Mill, below the village and set back near the Bell Hotel, survives with some of its earlier timbers in a rebuilding of *c.* 1866. Though the pond remains, the water-wheel was removed with most of the plant after the last war.

One of the few mills today in the London countryside, whose overshot water-wheel works the plant which still produces meal, can be the delightful surprise of Ford End, Ivinghoe, just west of the village. Also, its late characterful owner, Mr Jellis, delighted to show it working. Picturesquely set, the building is only average, but its interior is uncluttered, less the apparatus that once filled all available space inside, and for this reason provides a readier demonstration. A brick marked 1783 gives the date of the lower storey and the date of 1795 on one of the timbers on the stone floor level suggests the age of the last rebuilding here. Ford End mill may represent that belonging to Pitstone Neyrnut Manor in 1346.

Another tributary rising out of the chalk below Whipsnade drove certain examples. The first is Dolittle Mill, even today a weird-looking affair at first glance. After the First World War, it ceased working and now looks somewhat forlorn. The pool has sunk to a trickle and the tail seems to have vanished into a tunnel. No doubt one of the mills credited to Totternhoe in Domesday lay here, and in 1610 Richard Buckmaster owned a mill, almost certainly on this site, and Dolittle Mill remained in the same family till recent times.

The lower part of this water-cum-windmill composition consists in the main of a sixteenth- to seventeenth-century structure wherein can be seen the water-wheel, some of the gears and the hurst. Some time in the mid-eighteenth century, the miller, obviously finding the water supply inadequate, had erected the windmill tower. This massive brick cone is built not on normal 'earth-bearing' foundations, but upon a wooden framework supplemented by a couple of cast iron columns. The danger today seems to lie less in this risky support, where strain would be expected to show, than in the growing plants that, with the frost, are disintegrating the top rings of brickwork.

A parallel stream flows past the Icknield Way, to fill a large pool once driving one of Edlesborough's mills. The surroundings are pleasant enough but the structure, derelict, almost in ruins, bereft it would seem, of all plant, is unsightly, and contrary to the usual effect, leaves the impression of being both desolate and uninviting. Next comes Bellows Mill, though last century it carried the name Dyers Mill, which is now an attractive residence. Though the mill is nineteenth-century in design, the miller's house was of greater age. A turbine replaced the water-wheel a fair time ago, but the more recent conversion has enabled some of the machinery, in itself cast iron, to be left, and remain workable.

TRIBUTARIES OF THE RIVER IVEL

A branch of the Ivel collects a tributary passing through Campton and Shefford, which has several head waters rising below the Barton Hills. On one of these can be found Barton Mill, just east of the village. Once standing like a sentinel over the surrounding

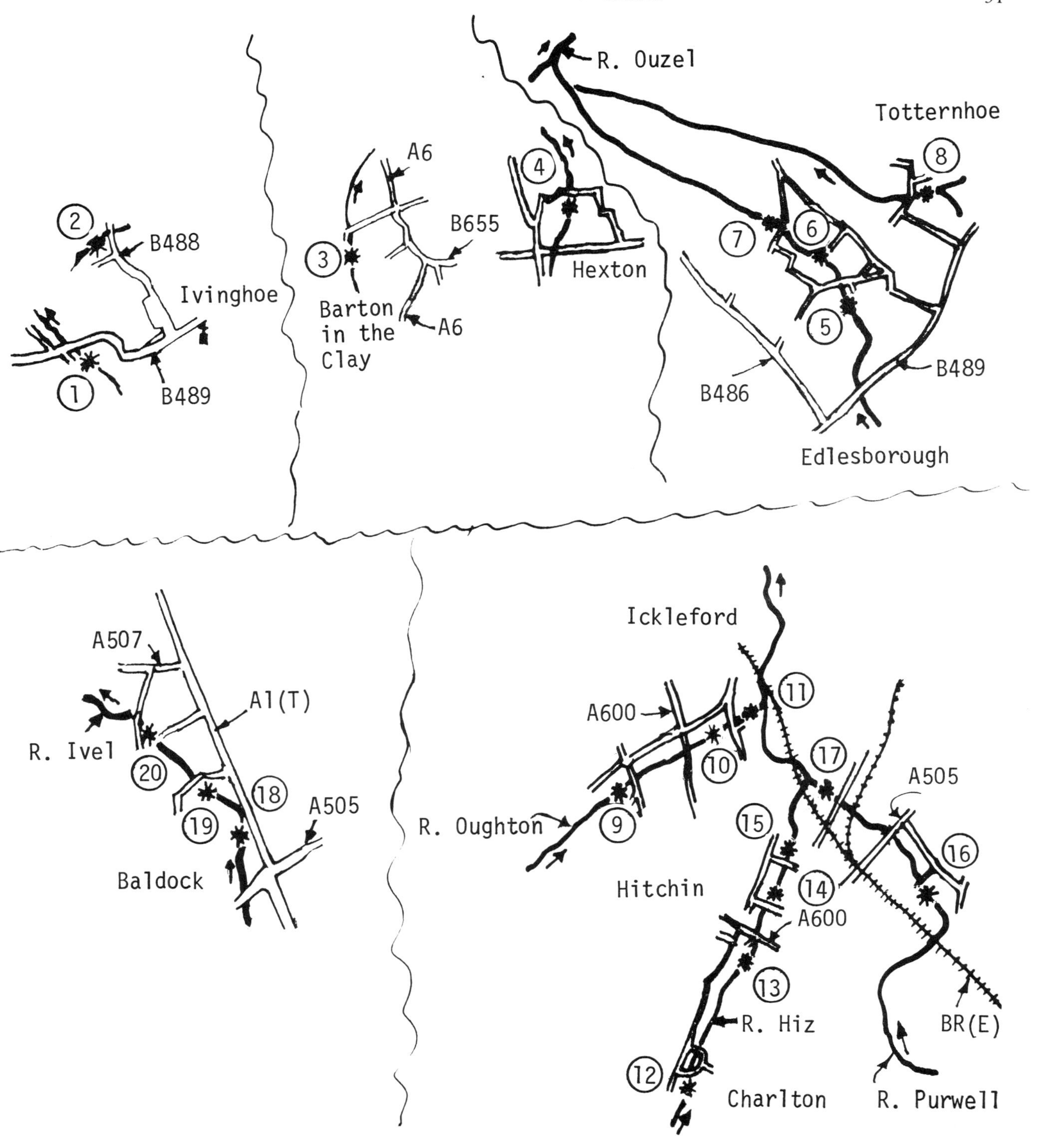

1 Pitstone Mill
2 Ford End Mill
3 Barton Mill
4 Pegsden Mill
5 Edlesborough Mill
6 Bellows Mill
7 Eaton Bray Mill
8 Dolittle Mill
9 West Mill
10 Upper Mill
11 Hyde Mill
12 Charlton Mill
13 Priory Mill
14 Porte Mill
15 Bancroft Mill
16 Purwell Mill
17 Grove Mill
18 Baldock Mill
19 Norton Mill
20 Radwell Mill

Tributaries of the River Ouse

Dolittle Mill, Dunstable, Bedfordshire (1930)

meadows and fields, today it does this somewhat sadly, the weather-boards no longer white, and the dilapidation far gone. Lack of lucrative trade closed it in 1926, the plant then being removed save for a water-wheel twelve feet in diameter on a wooden shaft. The present building replaced an earlier one of *c.* 1820, of which the sixteenth-century miller's cottage adjoining would have formed a relic. A predecessor belonged to Ramsey Abbey from 1255, when a moiety was purchased from Robert Peverel by Abbot Hugh of Sulgrave and held until the dissolution. Other proprietary portions are mentioned and, as late as 1611, one was held by Felix Wilson and was then stated to have come out of the one-time abbey mill.

Higham Gobion's watermill has vanished as completely as has its village. Records state that in 1279 Richard de Gobion and his mother, Maud, quarrelled over the proceeds, and at Richard's death in 1300 the inquisitor valued this mill at 13*s*. 4*d*. It also functioned as soke mill for the nearby manor of Streatley, both this last and Higham Gobion coming under the barony of Bedford.

Pegsden Mill at the outskirts of Hexton Park seems to have been transferred from Shillington in the thirteenth century. In the early thirties its descendant could be found — an odd-looking specimen of early nineteenth-century date, rendered and 'castellated', covered in ivy; a bit of a caricature, in fact. Today its condition is semi-ruinous, and inside can be seen the reason for its small size and, in a sense, its existence. Here the hinged bucket type of water-wheel sufficed as the prime mover for the pumps that filled Hexton Pond and supplied Hexton Hall. This building would occupy the site of a corn-mill, destroyed no doubt when the parkland landscape was being transformed in the early nineteenth century. Further north in the parish stood another mill, whose ancestor Domesday records as being destroyed, and which a charter of Henry I refers to as being rebuilt.

RIVER OUGHTON

One tributary of the Hiz rises to the west of Hitchin and, called the Oughton, used to drive two watermills, the first being West Mill in Pirton. Here at Domesday, it was presented in the eleventh century to the Priory of Hertford by Alan de Limesi, who was perhaps a brother of that priory's founder, Ralph de

Purwell Mill, Hitchin (1931)

Barton Mill, Bedfordshire (1931)

West Mill, Hitchin (1931)

Limesi. At a subsequent date Wiscard Ledet, lord of Ramerick, quarrelled with the prior over the rights of the monks to a certain 5*s*. that they obtained from tithes upon the mill and which should have gone to Wiscard's chapel in Ramerick (a later house still survives close to the defunct railway line near Henlow Station). A compromise was reached upon this dispute, and the monks agreed to supply a chaplain for celebrating mass in the chapel on four days a week, for which they received an extra 1*s*. 8*d*. in addition to the said 5*s*.

Manorial records describe the mill as being, in 1353, in so poor a repair that no tenant would rent it. However, this state must have eventually been made good as the last mill here had a sixteenth- to seventeenth-century timber frame behind the eighteenth-century brickwork, weather-boarding and slate roof. All that can be seen surviving from demolitions carried out in 1960 are some mill-stones, the lower walls of the watermill and a concrete spill-way on the position of the water-wheel.

A brick bearing the legend 'MW 1755', has been kept, and though the disappearance of the mill detracts from the appealing group — mill, miller's house and barns — the latter two are still there.

Next the Oughton stream flows to Upper Mill, but its function there is today more or less decorative, as the pool has been filled in. The nucleus of the present roller-mill establishment dates from about 1860, still retaining, faintly perhaps, a certain shapeliness and elegance. Both older mill and surroundings cannot escape the less pleasant visual dominance of the factory. The stream flows, neatly contained and 'land-scaped', to the bridge, then under, to join the Hiz.

RIVER HIZ

Charlton Mill, the first of Hitchin's royal manor mills on this river, ground coarse meal for animals between the wars, but only a portion of this Victorian building remains today. The cast iron water-wheel with steel buckets exists in position and the fine pool remains upstream. Charlton Mill, recorded in Domesday, and mentioned again in 1177, was the scene of a tragedy in 1279. That year 'John le Saretur went to turn the wheel of the mill of Reginald de Hiche when the wheels moved and drew him between cogs, whereof he died'.[1] As was the custom, the hamlet was declared 'in mercy' and probably would have been fined for allowing the accident, while for purposes of justice, the offending article, the wheel, was valued at 16*s*. 6*d*.[2]

Charlton Mill was involved in 1670 in a lawsuit, while six years later in common with another mill in the town it was considered too 'old and ruinated to do their work' and therefore was rebuilt. The mill-house, a pleasant building of last century, was saved from the ravages of the fire that in 1887 consumed the mill. Boards and corrugated iron were chiefly used in the rebuilding, a hardly inspiring selection of materials.

Next the Hiz flows through Hitchin town and drove the four mills that sufficed for the town. Edward II in 1317 founded a house of White Friars at Hitchin, part of which still survives in the eighteenth-century house, The Priory, and to the north-west of the precincts stood 'le malt milne'. The dam to its pool was repaired in 1471. The Ferrers and Philip couple were its possessors in the early part of the seventeenth century, and as a two-storeyed building with an external water-wheel and a cupola on a gabled roof, it survived until its demolition in 1806.[3]

The Hiz flows onwards and, in a canalised form, passes east of the churchyard, through an area that at a visit in 1931 came as a shock, so out of character did it seem in comparison with older Hitchin; the result of a wholesale clearance designed to redeem a quarter whose sluminess had been much publicized. No surviving watermill suffered, but the obliteration of part of the old town and its replacement, in layout and details both trivial and rootless, has barely improved in the passage of years. A less discredited technique would have spared much for later conservation.

By Portemill Lane the Hiz, by a devious system of falls, disappears under the roadway. The name of this thoroughfare comes from one of the king's mills that stood on the north side. It was repaired in 1471 and in 1677 the miller, one Edward Prier, was called before the manor court for omitting to scour the watercourse in front of his door — presumably his house adjoined the mill. The Crown disposed of Porte Mill by auction in 1813, and though rebuilt two years afterwards, it was finally demolished in 1852.

Old maps[4] suggest that another watermill was situated below the Almshouses in the Bancroft. It probably adjoined a long building that later did service as a leather-dressing works but, unless it could be identified with the mill belonging to William de Lindlegh and his son in the thirteenth century, it has a slender history. The Hiz joins up with the Purwell just below Grove Mill, yet to be mentioned, while the next mill to be supplied, could be found at Hyde Mill, Ickleford. Alas, it suffered demolition some years ago.

In 1931, Hyde Mill, then one of the oldest in Hertfordshire, rickety, yet still producing coarse cereals, consisted of a house adjoining a watermill dating in part from the fifteenth century or earlier, with a later wing placed at right angles. Though the surrounding meadows had not remained unspoilt, the group looked picturesque in all views, particularly when seen from the bridge across the dark reflecting surface of the pool. Unfortunately the larger part of the mill-house, which extended on the left side of the mill, had been demolished. An old bridge, looking as if it would collapse under the slightest weight, spanned the tail and connected the mill-house to the pathway. The taller centre portion, with a gambrel roof and the shed over the leat, were later additions. The Priory of Wymondsley possessed this mill in the middle ages and it seems to have been permanently leased to the Priory

Hyde Mill, Ickleford, Hertfordshire (1931)

Burnt Mill, Hitchin (1931)

of Elstow which, in 1354, was ordered by court to repair a runious bridge, probably a predecessor of the one shown. The Crown laid hands on it with the rest of the priory's property at the Dissolution, and in 1542 granted it to James Needham, surveyor, and clerk of works to Henry VIII.

RIVER PURWELL

A tributary of the Hiz is the River Purwell, rising in St Ippolitts. Purwell Mill in Great Wymondley, the first, ceased working long ago, became a milk distributing centre between the wars, and is today a store and offices. The mill building, *c.* 1850, remains a shell only, the sole interest now being the four queen post trusses carrying the roof and also the bin floor suspended by steel rods from trussed tie beams. The adjoining Georgian miller's house merits a glance but memories of a visit in 1930 confirm that the remoteness of the situation and the tidiness of the surroundings have suffered a decline.

Next the Purwell Stream drove Grove Mill. This particular example looked just as unprepossessing in 1931 as it does today, but the pleasant Elizabethan cottages and gardens sloping down to the stream are missing as are other more eyeworthy accompaniments. The tail seems little better than a rubbish dump and the mill itself, now a store and office, is unlikely to have retained any of its plant, including a small water-wheel then visible. This represents the second of the King's Mills of Hitchin and until the eighteenth century was named variously 'Shackling', 'le Shotting Mill', 'Shotlyng Mylle' and 'Shiekling Mill'. Renovated in 1471 and rebuilt in 1676 and then burned down the following year, it got the name of 'Burnt Mill'. A plant for silk manufacture was erected in it in 1849, but the scheme failed. Burnt again for the third time in 1889, the rebuilding that followed brought forth the present grim-looking edifice.

RIVER IVEL

This stream rises by Baldock. The first mill, now vanished, stood close to a weir at Blackhorse Farm in Bygrave, near an inn of the same name, where it is reputed Dick Turpin lodged upon one of his nefarious excursions. Greater fame, however, belonged to this mill through its miller's noted daughter, Mary, born here in 1726. A curate, unknown but probably of the parish, wrote the following poem. However doggerel-like, it was set to music, sold in the streets of London and became a popular song:

> Who has e'er been at Baldock, must needs know the mill,
> With the sign of the horse, at the foot of the hill,
> Where the grave, and the gay, the clown, and the lean,
> Without all distinction promiscuously go.
>
> This man of the mill has a daughter so fair;
> Of so pleasing a shape, and so winning an air;
> That once on the ever-green bank as she stood,
> I could swear it had been Venus, just sprung from the flood.

and so on.

Her fame sprang far and wide, pictures of the mill were sold, but the poor girl apparently suffered the discomforts of a publicity that seems to have been as insensitive as more modern modes. Perceval Stockdale, the amateur poet, recruiting at Biggleswade in 1757, rode over to talk with, or as it would be phrased today, interview her, for a wager. Many came from as far as London. She married one Henry Leonard who, after her death in 1769, married again and lived until 1802.[5] The rector wrote in the burial register 'Mary Leonard, wife of Henry Leonard' and added a footnote that 'this was the celebrated lass of Baldock Mill'.

Next can be seen Norton Mill, or what remains of it. The pool still beckons but alteration and demolitions, and much spoil left around, detract from what is still

Hexton Mill, Bedfordshire (1931)

an attractive scene. The watermill is part Victorian with a lucam, the other portion being a much altered vestige of a sixteenth-century frame. A turbine survives, but the plant has been removed. Below Norton Mill the stream forms the pool for Radwell Mill, which like the last belonged to St Albans Abbey. Radwell Mill can today be recognized in a residence, the conversion having been made many years ago. Nor are there indications that any of its mechanical parts would have survived because tenderness for such relics derives from a more recent exchange of sentiment.

NOTES AND REFERENCES

1 Hine, R., *History of Hitchin*, 2 vols., Allen and Unwin, 1927–9

2 Hine, op. cit., vol. I, p. 48

3 Chauncy, *Historical Antiquities of Hertfordshire*, Ben Griffin, London, 1700

4 Salmon, *History of Hertfordshire*, 1728; Clutterbuck, *Nat. History of Hertfordshire*, 1815; Bowen, *Map of Hertfordshire*, 1787

5 Hocking, S. K., (ed.), *Temple Bar Magazine*, vol. 58, p. 21

Chapter 15

THE COLNE VALLEY

THE RIVER MISBOURNE

THE Colne basin has its share of tributaries, the important ones joining from the western side. The first of these is the Misbourne, rising at Missenden, which once drove Deep Mill, near where the Great Central railway line spans the roadway. The abbey of Missenden, which is represented by a residence a quarter of a mile away, held this mill from its foundation in 1133 until the Dissolution, when it fell to various lessees, among them being, in 1584, the miller Anthony Nyxe. Afterwards a William Fleetwood purchased it and he and his heirs must have profited by the investment for they acquired one part of the manor of Missenden in 1668 and the other in 1684. Mill and manor stayed in the male line until 1745, and with their relatives until 1787. Only a portion of the mill erected during this family's possession remained between the wars, but sufficient to show how attractive the original must have been. The railway embankment carved the mill pool in two, and today no relic survives of the building. The Misbourne travels further down its open valley, until it fills the pool that once drove Little Missenden mill. This example had ceased grinding well before the last war and functioned as a dairy, but now remains as a most attractive dwelling. It consists of the common sixteenth- and seventeenth-century structure, almost rebuilt at subsequent periods. However, what attracts is the excellent way the machinery, i.e. the pit-wheel, spur-wheel, stone nuts, and the two mill-stones with their tuns, a crown-wheel, etc., has been incorporated into the entrance hall. In the hurst, which is below, the side has been glazed to allow sight of the pit-wheel and wallower and the remainder upstairs on the first floor includes an early type trundle wheel, altogether forming a most praiseworthy piece of conversion.

The Misbourne reaches Amersham, whose mills occur regularly in the records of the town, but to identify each existing one with a mill specifically mentioned in the past records is not easy. Town Mill, with its superb late seventeenth-century miller's house, set back a little and opposite Little Shardeloes manor house, constitutes an essential item in that 'survival' which is Amersham's market town street. One can remember Town Mill before the First World War and, though now no longer the same, it has by careful alteration become an attractive dwelling. The half timber frame would be of sixteenth-century origin mostly, but even while still a mill the interior had been rebuilt and altered considerably, as dates of 1700 and 1701 cut on beams inside testified.

Before the war a high banked pool, used as a swimming pool, could be readily seen from the lane, but now a high wall prevents this. The records say that a new water-wheel was fitted *c.* 1730, accompanied probably by a heightening of the pool banks.

This mill's remote predecessor, according to Domesday, was granted to one Roger, a tenant of the bishop of Bayeux. However, it became connected with the honour of Leicester, ground corn in 1419, malt in 1504, and in the early sixteenth century became the property of Robert Cheyne, a wealthy man thought to hold heretical opinions like others in this town. In 1772 it was purchased by the Weller family and afterwards used in connection with the now defunct Amersham brewery.

Bury Mill stands at the south extremity of the town. It closed down in the thirties and stood empty for a while before its conversion into a restaurant, an early case of a change now common. Before that time the plant, including the water-wheel, had been removed, though the stream tumbling over the race is visible from the inside. In the cleared space both shape and details proclaimed the mill to be seventeenth century with the usual incorporations of earlier parts, particularly the wall between mill and the former miller's house. The present use required removal of much of the 'stone' floor and an array of grain bins but the change of use has undoubtedly allowed the building to remain.

River Colne I

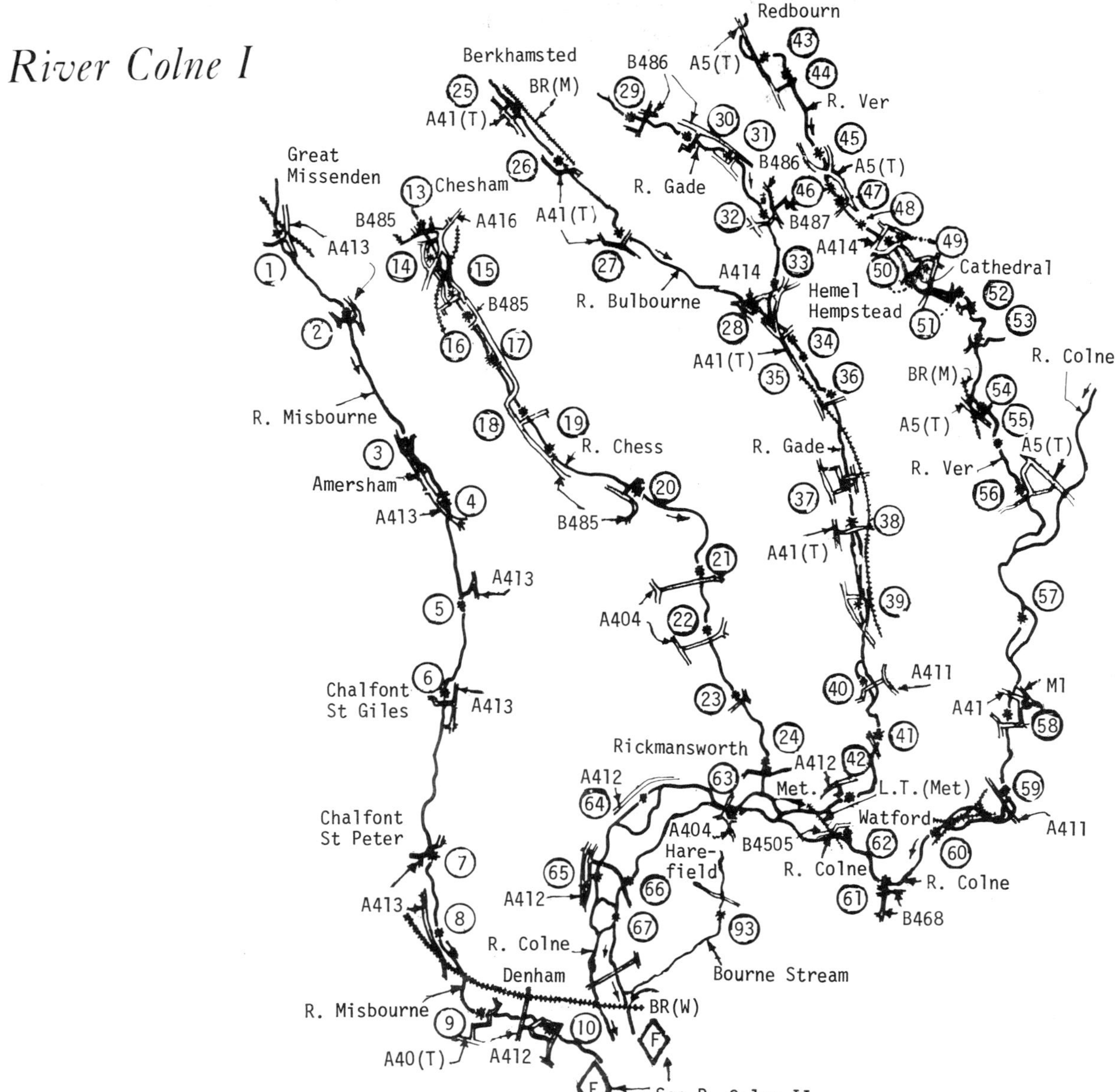

1 Deepmill
2 Little Missenden Mill
3 Town Mill
4 Bury Mill
5 Quarrendon Mill
6 Chalfont Mill
7 Chalfont St Peters Mill
8 Oak End Mill
9 Moor House Mill
10 Town Mill
11/12 see list for River Colne II
13 Bury Mill
14 Amies Mill
15 Lords Mill
16 Cannon Mill
17 Weirhouse Mill
18 Blackwell Mill
19 Bois Mill
20 Chenies Mill
21 Sarratt Mill
22 Solebridge Mill
23 Loudwater Mill
24 Scotts Bridge Mill
25 Upper Mill
26 Lower Mill
27 Bourne End Mill
28 Two Waters Mill
29 mill site (Water End)
30 Noakes Mill
31 Piccotts End Mill
32 Bury Mill
33 Albion Mill
34 Frogmore Mill
35 Apsley Mill
36 Nash Mill
37 Home Park Mill
38 Langley Mill
39 Hunton Mill
40 Grove Mill
41 Cassiobury Mill
42 Croxley Green Mill
43 Dolittle Mill
44 Redbournbury Mill
45 Shafford Mill
46 Butlers Mill
47 Prae (New) Mill
48 mill site
49 Kingsbury Mill
50 Silk Mill
51 Cotton Mill
52 Sopwell Mill
53 New Barnes Mill
54 Berry Dell Mill
55 mill site
56 Moor Mill
57 Aldenham Mill
58 Bushey Mill
59 Watford Mill
60 Rookery Mill
61 Hamper Mills
62 Batchworth Mill
63 Rickmansworth Mill
64 Mill End Mill
65 Troy Mill
66 Harefield Mill
67 Jacks Mill
93 mill site

Town Mill, Amersham (1929-31)

Misbourne Mill, Chalfont St Giles (1931)

At Domesday, Ulviet held this mill from High de Bolbec, and in the late thirteenth century it belonged to Wedon Hill Manor. Ralph de Wedon afterwards gave this example, then bearing the high-sounding name of 'Brightriches', to Missenden Abbey. Evidently another manor called Tomlyns had a share in its profits for a century, but ultimately it came with Wedon Hill Manor to the possession of the Drakes of Shardeloes, once the chief landowners in Amersham. Though a corn-mill, and for most of its life it was used as that, in 1825 it was used in connection with the manufacture of silk.[1]

Quarrendon Mill stopped working soon after the war of 1914-18. It lies in the valley, still pleasant and secluded in its situation. Though the outside walls, miller's house and mill are one, with an obvious Victorian date, the shape at first suggests an older structure underneath. However, this is not so. It is all of one building, including the dam, the pool (now dry — the wheel race giving a twelve-foot fall) and the walls of the tail. Apparently the cast iron plant went for scrap in the last war. The Domesday mill on this spot was held by Gozelin le Breton and later passed to the Quarrendons, a long defunct Amersham family, their name to be perpetuated by the mill. In the thirteenth century, it reverted to the chief manor and by the seventeenth century had come to the Drake family.

Chalfont Mill as late as the thirties stood in a picturesque position, backed by a wood of young poplar trees, and with an attractive footbridge and watersplash at its extremity. Its predecessor, mentioned in 1247, was in ruins by 1349, probably as a result of the Black Death. It once comprised a long run of buildings, all of which were old, except the chimney and factory-like addition at the south end. The plant ceased to be used long before the last war, and the mill together with the miller's house adjoining had become one house. Today, it would appear, further alterations are taking place. The pool head has been shifted upstream and the overflow diverted, and it is now difficult to locate the position of the inside (and vanished) waterwheel. The group was in the main of sixteenth-century building, with older pieces of oak evident in its make-up. The wooden cog-wheels, visible in 1930, can no longer be seen today for the later conversion had all these removed. Likewise, one wonders what happened to the panelled plaster ceiling and the seventeenth-century wall painting, once to be seen inside the miller's house.

The mill belonging to Chalfont St Peter stood on the north of the village, but as late as the thirties no remains were visible, though it was shown in 1825 as a silk mill.[2] The earliest mill here was given with other lands to Missenden Abbey, *c.* 1230.

Next on the Misbourne, by Chalfont Park, once stood Oak End Mill. This mill, together with some land, belonged to the Templars, afterwards to the Knights Hospitallers, as subsidiary to their estate in Hedgerley. The pattern of the banks on the watercourse at Moor House Farm, Denham, indicate a mill site. If so, when it vanished is not recorded. Between the years 1216-72 the name of 'Robert of the Mill' occurs three times and the mill implied could have been situated at Moor House Farm. A bridge stood near this example (then called 'Brokenmyln') when the Abbot of Westminster was warned by the Denham Manor Court that failing reparations before the next court day, he would incur a fine of one shilling.

The Abbey of Westminster held Denham Manor in medieval times and consequently the Abbot as overlord was always referred to as the owner of the mills, one of which was situated by the village, and the other three on the Colne near Uxbridge. Town Mill, down the lane from the church, stood for several years disused, but much alteration has taken place including the removal of the roof and the plant inside.

The original building dating from 1857, was of a restrained design and an acceptable element in so attractive a village. No doubt when the alternative was demolition, it was seen that the shell could be made into a dwelling to meet a demand in this area so near to London, a fortunate rescue, though the mill of 1929 can scarcely be recognized in what stands there now.

Someone at the time added on a new (south) wing and planted on many of those frippery details so beloved of the speculative builder, the general effect being far removed from that natural comeliness which used to belong to a working mill of this kind. A predecessor, called the 'Corn Mill of Denham', was valueless in 1377 because no tenant would rent it on account of its bad repair. However, full accounts of the partial rebuilding have been preserved (cf. Vol. I, p. 56). Names of many lessors survive: in 1391 Walter Holtyng; in 1401 Richard at Terne the miller; in 1440 Hugh Bereacre holding the same position; in 1441 it was John Lythting; then in 1450 William Bereacre, possibly a relation of Hugh.[3]

RIVER CHESS

This river drove many mills, none of which still work at their original trade, though some of the buildings have been kept. Another survival, a kind also becoming rarer as time passes, can be found at Chesham. This consists of those channels, originally made for supplying the mill-pools, which weave their way in and along the old lanes, the latter relatively intact and full of houses representing many different periods, a welcome contrast to the destruction elsewhere in this town.

The first example seems to have been Bury Mill, destroyed a fair time ago. It stood below the church in Church Street, the pool remaining but modern houses occupying the site. Though not a Domesday recorded mill, it would represent the Cheshambury Manor Mill held in the early twelfth century by Richard Sifrewast who bestowed it upon that collector of watermills, Missenden Abbey.

At the corner of the Amersham Road and Amies Lane a much altered eighteenth-century mill-house indicates Amies Mill. Though the latter disappeared long ago, the banks of the pool and the race remain. The De Broc family held Payne Mill in 1275, which they leased to the same Missenden Abbey, while in 1286 their possessions included a corn- and fulling-mill, probably on the same site if not within the same building.

The next is Lord's Mill, the chief manorial mill for Chesham. Mentioned in Domesday, records refer to it in 1312, 1481 and in 1580 when two water-wheels drove its machinery. The present mill, now much altered, patched and rebuilt, dated from the early seventeenth century. Some of the original walls of thin bricks remain; the chimney stack is certainly a good specimen of that type of detail. Equally ancient is the miller's house adjoining. The gables of the house have their original timber framing, probably of the same date as the mill, but much of the external walling was rendered over last century.

Little of the former plant remains inside, while the outside is now immobilized, as the grist trade ceased after the last war, leaving now only a small amount of animal meal being produced. The deteriorating repair of the building bodes ill for its future. The position of Lord's Mill, at the head of a lade, the latter running alongside the road, resembles others in this valley, but the arrangement of the wheels has changed, and clues to this story are somewhat confused. It would seem that the present wheel drove the plant on the north side, the original position of the race with probably two wheels side by side at an earlier date. About 1880 the miller installed a wheel at the rear, a large and interesting one to judge from a photograph in the office of Mr Oakin, the tenant, which he shows to visitors. The centre wheel was removed and subsequently the wheel arrangement reverted to the pre-1880 plan.

The next downstream, Cannon Mill, presented a pleasant feature in the twenties, typical of the Chess Valley — a structure of two weather-boarded storeys and a tile roof standing at the head of an ample pool. Before the last war it had been drastically renovated, but by 1956 it was in partial ruins and today only the position of the race remains; it is certainly not possible now to see the line shaft type of gearing, a kind unfamiliar in this area.

Weirhouse Mill (building only) stands much in evidence, with its miller's house, and the supporting barns that accompany it, still there. The meadows once surrounding it on the south side have given way to a factory. The proprietors have retained, utilized and renovated most of the original grouping, for which they should be praised. The watermill has lost its wheel and also that portion of the building which housed it. Weirhouse Mill was mainly early eighteenth century following the common idiom — weather-boarding on brick lower storey and a tile roof.

Further downstream Blackwell Mill stood at the weir near Blackwell farm. It disappeared last century, but was, it seems, of medieval foundation, being mentioned several times in those centuries. Another destroyed example — Bois Mill — stood near the road, the miller's house apparently surviving near the spillways. Of Domesday origin, a predecessor belonged to Chesham Bois manor in 1310. Francis Cheyne as Lord of the manor held the mill for a lengthy period ending in 1645 but in 1585 he quarrelled with Henry Gordon, evidently because the latter had tried to stop custom coming to Cheyne's mill. Gordon, 'a disorderous and contentious person', was charged with obstructing a laden horse accompanied by a man, whom he 'had stroken and verie evillie abused'. A few years later the mill was working at another trade, as witness this entry for 25 July 1592 in the parish burial register: 'John Carter of the fullynge myll', and 18 September of the same year saw the burial of 'the wife of one that had taken the fullyng myll of William Carter'.

The next example on the Chess can be found by the bridge below Chenies village, or Isenhampton as it was once more melodiously called. Mentioned in 1323 as a fulling-mill producing a rent of £3. 0*s*. 0*d*. p.a., it later became a corn-mill and in about 1800 a paper-mill. Much of the latter building still survived until the recent demolitions. The remainder has been made into a residence, successfully enough no doubt internally, but outside, the effect commands much less admiration. Like much of this restoration during the last decade, the saving of an old building has been accompanied by certain 'architectural embellishments' that have almost taken away the character which it was obviously intended to keep.

The overshot cast iron water-wheel has been kept as a feature and the leat also. Lots of 'rocks' have been placed to form a garden on the site of part of the mill, attractive enough in itself but hardly appropriate. One of the cast iron gear-wheels finds itself built into a wall to form a window opening. Using discarded millstones for steps and pavings may be acceptable, but is this choice any better than a gimmick? For so near to London the surroundings of Chenies Mill leave little to be desired — the pool, the spillway or tumbling bay, the overflow channelled within banks of grass, not of concrete, and framed in meadows.

The Chess flows on, passing three sites only — Sarratt Mill, and then Solesbridge Mill near Loudwater, and a third of the same name, to drive, or once drive, the last mill on its length. Even at this stage the stream keeps its sparkle and clearness (surprisingly, considering the changes in the valley), as can be seen in the waterways at the rear of Scotts Bridge Mill near Rickmansworth. This mill ceased to be such long ago, and today suffices as part of a store for M.G.M. Apart from the disused mill-race, and the unmistakable dimensions and shape, one would hardly believe in the building's origin. The core, of the early nineteenth

Chenies Mill, Buckinghamshire (1931)

century, incorporating pieces from an older structure, has kept nothing of the plant, nor have the two mid-nineteenth-century extensions at either end.

RIVER BULBOURNE

This river rising between Tring and Northchurch joins the Gade at Box Moor but before that come the sites of three mills and the substance of another. Berkhamsted possessed two mills at least. A large and attractive building stood on the east side of Water Lane (originally Mill Lane) and can be identified with the 'Sisethmulle' of 1357 and the Upper Mill of the sixteenth to seventeenth centuries. It is also, probably, the mill whose water-wheels worked the pumps to supply water to Berkhamsted House in the seventeenth century and which by 1650 lay broken and useless. The great castle nearby had its own mill, either adjoining the one just mentioned or perhaps driven by the water from the castle moat. Not being mentioned again it must have disappeared following the disuse of the castle. Half a mile out of town stood the shell of one placed in the turn of the lane. Here must have stood the 'Bankmulle' of 1357 and the Lower Mill of later centuries.

Berkhamsted St Mary (or Northchurch) had a mill at Bourne End in a detached portion of the parish near the boundary and one occasionally included in records of those of Hemel Hempstead. It was called 'Welpesbourne milne' and was held in 1290 by Edmund of Langley's College of Bon Hommes of Ashridge, and though mentioned as 'Burnmyll' in 1540 it still retained the name Welpesbourne Mill in 1609. Sometimes Westbourne Mill was the appellation bestowed upon it. Perhaps the name derived from the same origin as Whelpley Hill near Chesham. Bourne End Mill, its most recent name, is much younger than its medieval site, with the building being architecturally somewhat indifferent mid-nineteenth century. Though it became a store long ago the water-wheel remains *in situ*, the hurst as well, though this is so barricaded with boxes and other storage impedimenta, that one could only just see the wooden spur-wheel, pit-wheel, the crown-wheel and the upright shaft.[4]

Hemel Hempstead boasted an important corn market, commented upon by eighteenth-century travellers. Hence its numerous watermills, that provoked William Ellis,[5] the agricultural writer, to remark 'that there are more watermills erected on the river Gade near this market than in any other in England in the same distance of ground'. Obviously he had never visited the Loose Valley near Maidstone, or the Pennines. These Hemel Hempstead mills numbered four at Domesday and until the mid-thirteenth century when they had increased to five, while in the sixteenth century, three of them housed two water-wheels each. The last on the river Bulborne, Two Water Mill, is now only a site. Its predecessor in the thirteenth century was bestowed, like the others in Hemel Hempstead, upon Ashridge College. In 1290 it was called 'Tuewater Milne' and in 1540 'Le Fullingmyle' and forty years later the property, containing two water-wheels, a house, garden and a fishery was held by John Henry and Edwin Waterhouse for the life

Noakes Mill, Gaddesden (1930)

of the survivor. As they still kept possession of the place in 1650, so long a tenancy must have meant a great age for both men.

RIVER GADE

Hertfordshire was once noted for numerous watermills; for instance, in 1595 forty examples could be found within ten miles of Hertford, and William Ellis, quoted previously, said in 1732 that 'they run upon this grain [wheat] as being a county best furnished with of any other with watermills for grinding the same'. Unfortunately most of those observed by Ellis in his travels have vanished. The first mill on the Gade, near the appropriately named Grist House Farm, is Noakes Mill, the last descendant of 'Okmyl' recorded as grinding corn in 1369. The half-timbered mill-house of a later mill, stands above, old enough to be the home of the miller, who had William Ellis for a customer. The latter wrote 'I now grind my wheat Grist with a reputable eminent miller at Oak Mill . . . I am satisfied of his Fair-dealing . . .'

Though today the picture is much different, fortunately this not unattractive Victorian building is being retained. In 1930 a Youth Hostel made use of the nearby converted stable building while the mill itself stood disused and with the plant partly dismantled. Since then another inviting feature, a ford, has succumbed. The plant has now been removed except the pit-wheel and the cast iron water-wheel. As the

Huntonbridge Mill, Kings Langley (1929)

Grove Mill, Watford (1929)

Park Mill, St Albans (1932)

Piccotts Mill, Hemel Hempstead (1929)

pool has been filled in, this wheel is unlikely to revolve again.

Further downstream, at the edge of the New Town of Hemel Hempstead, Piccotts Mill can hardly be overlooked. A late Georgian mill-house shows its importance at that date while the mill itself is about 1860. The building and site alone now claim interest, for the old water-wheel and plant have been replaced. This is also an example highlighting the problem of keeping this sort of building, which the local authority would wish to be kept, less for itself than for a contribution to the amenities, amid so much new townscape. The eventual cost of renovating so bulky an edifice will probably be more than such an establishment is prepared to meet.

In 1086 the Mill was called 'Picott', in 1290 'Picotes Milne' and at the Dissolution 'Piggottes Mill' but not content with such spelling acrobatics, it became in 1580 'Pickets End Mill' and it then consisted of two watermills for grinding corn.

Bury Mill shared fortunes with Bury House nearby. It was called the Court Mill in 1268, Burmille in 1290, and by the Dissolution had attained its present name. Rebuilt, it would seem in the nineteenth century, its shell could be seen as late as 1935 amid several older cottages. All this has changed, the building has been removed, and the pool and waterways have been transformed into a most commendable piece of landscape design. In its way, a net improvement. When Hemel Hempstead New Town started its widespread alterations to the local topography, the nineteenth-century Albion Mill site at More End near the railway embankment must have succumbed.

The watermill site at Frogmore has been replaced by a factory. Though named 'Fellingmilne' in 1290 it was called at the Dissolution, 'le Convent Myll' and retained the same name in 1580. Like Piccotts Mill and Two Water Mill it belonged to that long-lived couple mentioned in connection with the Two Water Mill (see above). Upon the site, in 1804, a machine of some significance, for manufacturing continuous rolls of paper, was erected by Henry Fourdrinier which must have been the mill and machine visited by Emperor Alexander of Russia, in 1814.

Being greatly interested in the manufacture, the latter arranged for the brothers Fourdrinier to set up a similar plant in Russia for £700 a year. But they never received a penny. This ill-starred pair suffered much and Parliament, after considering their case, voted £7,000 to be given to them. Contemporaries thought this inadequate and an additional subscription was raised, mainly by the paper trade, and one of the brothers, Henry, lived comfortably in retirement until the age of eighty-eight.

The next three mills have long been replaced by factories — Apsley Mill and Nash Mill — the latter deriving from the 'Ash Mill' or 'Nayshmill' of medieval records, the first for corn and the second for fulling and the third mill, lying below King's Langley Church, would represent the medieval 'Queene myll' of the middle ages.

The two mills at Abbots Langley generally followed the fortunes of the manor, the overlord being the Abbot of St Albans. The northernmost, Home Park Mill, is also a paper factory but upon its site the 'Molyn du Petite L'Angele' was leased by Eleanor, Edward the First's queen. She paid the rent regularly to the abbot but the next queen, Isobella, who continued with the lease, had obtained effective control of the country by 1310 and among her numerous misdeeds she proceeded to ignore all her landlord's rights. The situation continued until 1335 when the abbot was determined to obtain his rightful rent, and also the rent unjustly withheld for twenty-four years, to be set against the abbey's debt to the Crown. The abbot won his case and the 'she wolf of

France', now the Queen Dowager in enforced retirement, had to pay £1 per annum for the mill in future. This mill and the next one downstream were rented to John De Chilterne in the fourteenth century. He is a man recorded as harbouring a venomous hatred towards the Abbot of St Albans. His quarrel, commencing in 1364 over this manor and mills, lasted for nearly thirty years.[6]

Hunton Bridge Mill, the second example, an early eighteenth-century building in the by-lanes surrounded with cottages and barns, represented an example obtained by Sir Richard Lee in 1544 and afterwards it came into the hands of the noted mill acquirers, Ferrers and Philip, at the beginning of the seventeenth century. Two hundred years later it was purchased by the Dickinson family, the well-known paper makers. Since the last war the place has changed much, only the lower storey walls from the demolished mill being kept to form a walled garden for the adjoining residence. Happily the cast iron overshot water-wheel was also kept and, seemingly fixed in position, acts as a spillway.

The river Gade flows towards Grove Park and once supplied Grove Mill. It has been disused since before the last war, the water-wheel being little more than a frame. The building itself, erected as the date panel announces in 1875, assimilates to its delightful surroundings, and no doubt it will be the last of a line dating back almost to Domesday.

Only the shell of Cassiobury Mill is likely to be found today. In 1265 Petronilla de Ameneville received a predecessor for life from Abbot John of St Albans. With other lands it was given to this lady for income, as she had taken the veil, and it was to endure as long as she kept to her vow. However, soon afterwards she married a John de Grava and the property reverted to the abbot, a renouncement of all claim upon its being made in 1271. Petronilla was grand-daughter of a previous lord of Oxhey Manor and before her death she founded a Chantry in Watford Church. Later on Cassiobury, or Cassio Mill as it was sometimes called, had the name 'Tolpade' and by 1364 it belonged to a certain John Aignel and had fallen into decay.[7]

Croxley Green paper factory marks the site of a Domesday Mill and one presented in the twelfth century to the nun's priory at Clerkenwell, by Richard de Croxley.

RIVER VER

The first on this stream could be found south of Redbourn alongside Watling Street, a rather pleasing little building of about 1800 with a fine pool on the other side of the road. It was called 'Dolittle Mill', a name which may at the time have described its state but was normally applied to mills that were the first on their respective streams. It cannot be seen today, so evidently road widening sealed its fate. The two water channels can still be made out but the once fine pond looks little better than a swamp. This example belonged to the Abbey of St Albans and was called 'Bettespol Mill' in the fourteenth century and 'Le Corne Mill' at the Dissolution. History has not forgotten it as it achieved a certain notoriety from an ordinary accident having an extraordinary end which took place in 1334.

A little girl of about five years old, doubtlessly playing by Bettespol Mill, fell into the water — either into the pool or into the leat. (As late as 1932 a shed covered the wheel (or turbine) but even then it was possible to observe how easy it would be for a child to fall in.) However, the wheel, according to the chronicler a powerful overshot, was revolving 'swiftly and surging through the water' and drew her body underneath. It was expected that she would be pulverized, drowned or killed but when the 'tremendous pressure' of the wheel threw her clear of the water she was free from mutilation but apparently dead.

When a crowd of neighbouring folk ran to the scene, the child's mother was 'lamenting vehemently' over the sufferings of her dead child, clasping its body and offering both prayers and pence to St Alban for his intervention. Whereupon the girl began to revive and in a short time showed herself completely restored to life.[8] Thomas Walsingham, who recorded this incident, may well have called it 'a wonderful deed performed from the intervention and efforts of St Alban'. He was greatly attached to the abbey and would forget nothing which reflected credit upon it and in that year, 1334, any incident calculated to raise its reputation in the eyes of its tenants was not to be left unexploited.

Another incident occurred in 1381 when the tenants of Redbourne showed no more love for their lord the abbot than did others of their neighbours. After several times threatening to level the bank which held back 'Bettespol' they carried out this threat, but the object, unless it was to render the mill useless, seems to have been obscure.

Next down the stream comes Redbournbury Mill, one whose rents augmented the finances of the chamberlain of St Albans Abbey. It is recorded that in the late thirteenth century the mill caught fire when a strong westerly wind was blowing. It was burnt to the ground but the chronicler added that, owing to the presence of numerous trees surrounding it, the neighbouring buildings escaped damage. The present building is not, of course, as old as a rebuilding after that catastrophe, most of it being about 1780, with very large portions from an earlier mill. Once, the surroundings were quite idyllic, but even now they are still pleasant. The water-wheel can still be seen and most of the plant remains and is interesting, in that it represents the less frequently met-with line shaft and has many old cogs and gear-wheels. Though working after the last war, that has now ceased.

The next, Shafford Mill, near the farm of the same name, is today little more than a pleasing-looking shell,

Dolittle Mill, Redbourn

Redbournbury Mill (1929)

Sopwell Mill, St Albans (1929)

Shafford Mill, St Albans (1929)

for work there ceased long ago. After that the water-wheel, which can still be seen, pumped a water supply to the mansion on the hill to the east, but it is now needed no longer. All the same, Shafford Mill, as a landscape feature, is well worth a look. The pond remains, the structure has been kept repaired and the immediate surroundings are not threatened as yet.

The Ver, flowing southwards after Shafford Bridge, was reputed to have driven a mill of which no signs remain today. A witness in an Exchequer lawsuit said that it had been built by a Londoner and, though at first the property of the Bacons of Verulam, it was owned later by William Preston who, at that date, held Kingsbury Mill in St Albans. In the late sixteenth century it was worked by a miller named Butler from which it took its name.

Next came Prae Mill. This mill was associated with the hospital and priory of St Mary De Pre or Prez founded in the twelfth century by Abbot Warin of St Albans and it took its name from the Praetorian Gate of Verulamium which of course lies across the nearby fields. As an establishment for leper nuns, it survived until 1527 when Wolsey earmarked its property for one of his projected colleges.

Today there seems nothing more than a couple of shed-like buildings near the weir. The foundation character of the hospital mentioned two mills the nuns could use: one was the Abbey Mill and the other 'Dichmulne'. The latter, which was destroyed before the fifteenth century, might have stood on this spot, or else upon a site of a medieval mill further down which will be mentioned. After the Dissolution, John and Francis Machell gave Praemill to the Bacons. The manor of Gorhambury, to which it was attached, was purchased by Nicholas Bacon, father of the more famous Francis, Lord Verulam. Sir Nicholas, a typical successful lawyer of the sixteenth century, bought the estate in 1550 and some time afterwards erected upon the site of Prae Mill, a machine to supply Gorhambury House with water — about a mile away with a hundred feet to rise: that was in 1563. When Sir Nicholas died in 1579 his widow, the Lady Ann, feeling no need for the water supply ordered the pumping apparatus to be converted into a mill. Evidently the local folk with the usual spontaneous way of peasants, gave it the name 'New Mill'. Lady Bacon leased it to one George Olebye who had previously held the Abbey Mills further down the Ver. Little time elapsed before this man tried to obtain the custom which legally belonged to the latter mills and through his success, a lawsuit at the Court of Exchequer followed. Mrs Olebye, who was called as a witness, was admitted to be 'an excellent wife' especially as she had been invaluable in obtaining business for her husband's mill, having approached some of the Abbey Mill's customers and 'persuaded with them' to patronize her husband's mill where the toll was much less.

As a part of the defence George Olebye declared that the demolition of the water pumping engine had resulted in a better supply of water to both mills lower down, namely Kingsbury and Abbey Mills. New Mill had been demolished by the time of the Napoleonic wars.

Between this last mill and Kingsbury Mill, in the year 1887, the remains of a medieval mill were unearthed in the river bed half a mile up from Kingsbury Mill but there seemed some doubt as to which one has this identity in the records.[9] Butler's Mill, mentioned previously, may have occupied this site.

Kingsbury Mill stands in its familiar place at the end of Fishpool Street, but is no longer working or workable. The picture is still attractive, though some of the surroundings have not been improved by the bypass to the north, or by the rash of spurious 'Regency' houses in the neighbouring field. Originally the property of the abbey, Kingsbury Mill is well recorded if such references could be easily unravelled. The Crown, its owners after the Dissolution, leased it out, and by the time of Elizabeth, it had come to William Preston who also held Prae Mill. John Oliver's map, *c.* 1700, shows a single building over the stream, a cartographer's simplification, because the two buildings, the plum coloured brick walls of the miller's cottage and the timber granary at the rear would have been standing then. A later map,[10] dated 1766, shows a complex of buildings which if not to be interpreted too literally, mark an increase in trade.

Inside, the plant follows more or less the standard pattern, mostly in iron; a nineteenth-century replacement of the machinery installed in the eighteenth-century rebuilding. The hurst framing looks like an adaptation; there are three sets of stones, the casing of one being covered with zinc; and the crown-wheel has slipped down the cast iron shaft. The internal water-wheel cannot now turn, but close at hand can be seen the pen-stocks and immediately adjacent, the overflow spillway, about two feet wide. It is to be hoped that St Michael's Mill, as it is sometimes called, will continue, as now, to remain a part of St Albans.

Through the public park, the Ver comes to the Silk Mills near the Fighting Cocks Inn. The mill buildings, in part, have been made into dwellings but the site could date from the foundation of the abbey. It would seem that the main monastic mill stood on the position of the building just mentioned. Some eighteenth-century additions provided further buildings in this complex, for the map of 1634 and John Oliver's map of 1700 show only two buildings side by side. The second mill, on the other channel, may be *c.* 1770 as explained by the following — 'as you enter St Albans a curious mill lately been erected for polishing diamonds, it is driven by water and saves a considerable expense as they used to be driven by horses or men'.[11]

How long this process lasted cannot be confirmed but corn-grinding had soon become the trade again, because in 1804 John Woollam acquired the place and installed therein silk-throwing machinery, hence the name. Buck's map of 1822 shows the Silk Mill as two simple buildings which indicates that its later extent was a nineteenth-century addition. Considerable alteration has now taken place, as can be seen from the public path which passes these mills. Some of the buildings were taken down and also a wall so that the spillway can be observed as it flows through a rock garden. The other buildings really comprised a nineteenth-century factory and the remaining structure has been converted, but the picture here is singularly unevocative of its story.[12]

One of the originals of these mills supplied the needs of the monastery and became the soke mill for monastic tenants in the growing borough. Between 1235 and 1260 Abbot John repaired these and probably the brewery which stood to the south-west. Thinking that the abbey ales did not set a good standard, he also set aside a large quantity of corn, i.e. barley for malt, hence the name 'Maltemyle' often mentioned with the Abbey Mills. Abbot Richard of Wallingford, among his many activities towards rehabilitating the abbey finances after the mismanagement of his predecessor Hugh of Eversdon, rebuilt the mill-dam and improved the meadowland between this mill and the chapel of St Germains in the nearby meadows. Around 1430 a new waterway or sluice gate was erected at the Malt Mill.

The great quarrel between the abbey and townsfolk over feudal rights and duties in general, and that of the mills in particular, however interesting, would make too long a diversion to be set out here.[13]

The Ver next comes to Cotton Mill, which survived as a Victorian building. The early twelfth-century scribes called it 'Stankwell' or 'Stankfeld' Mill, a name derived not from its condition but from 'stank', meaning pond. As a 'Fullerstrete' existed off Hollywell Hill leading to this mill, presumably it was the 'fulling mill below Eywood' which belonged to the abbey. In 1247 a certain Robert Stanhard was convicted of stealing woollen cloth from this building, while after the great storm of 1362 the abbot ordered the rebuilding of the 'Molendinium de Stangfeld' in order to give the townsmen no excuse for taking their cloth elsewhere.

After the Dissolution it became a paper-mill and later fell to Sir Richard Lee who, apart from possessing many watermills in Hertfordshire, also obtained a major portion of the dissolved abbey property.

The next is Sopwell Mill, where the lade and overflow channels make a moat around mill and house. Although long closed down, the group still makes a pleasing picture and its water-wheel has long been familiar to users of the neighbouring golf course. Though the site is ancient the mill now to be seen dates from 1868, after a fire had consumed its predecessor. It has been suffered to remain little altered since its disuse after the First World War. The gearing remains, with two of the three sets of stones, though the usual ancillary apparatus, bins and hoist gearing, have been removed.

The original designer of Sopwell Mill employed in its construction, uncommon for the time, rolled steel joists, yet in the hurst structure used pitch pine (twelve inches by twelve inches and twelve inches by six inches), probably because this material absorbed vibration. The gearing, in cast iron, has the pit-wheel engaging a cog on a line shaft from which other cogs drove the three spindles, one of the latter being geared to turn fastest, presumably for barley meal. The water-wheel decays away, but is not yet beyond reclaim. Here can be seen a cast iron axle, spokes and frames, with, morticed into the latter, wooden spars to carry sheet iron buckets.

The next, New Barns Mill, was, until the last war, a modern flour (roller) mill. Today, it is an engineering works, the plant being displaced long ago. Named Cowley or Sopwell Mill as early as the twelfth century, it belonged like the others to St Albans Abbey and also like them, afterwards came to Sir Richard Lee. Road widening has struck at Berry Dell Mill, the watermill following, at Park Street. Part of the mill has been lopped off in the alignment, less destructive certainly than the loss of those many cottages, which thirty years ago, made the village well worth a second look. The lade is rubbish filled, and what was the mill seemingly acts as a store, a sorry end to the last descendant of the 'Le Parkemulle' of the twelfth century. Records say that Abbot Richard rebuilt a predecessor here in the mid-fourteenth century, with another rebuilding in 1397 at a cost of £22. No relic of these could be found in the present nineteenth-century structure.

Lower downstream Chauncy[14] depicts a mill, and if it ever left any indications, the recent excavations for gravel would have erased them. Similar visual turmoil has greatly reduced the one-time beauty that its surroundings gave to the next (and last on the Ver) — Moor Mill. Mill and miller's house with accompanying barns survive to delight the eye and its interior will reveal many items of great interest.

Moor Mill worked at one time in conjunction with New Barns Mill upstream, but closed down just after the last war. Though its present function is no more useful than as a store, its owner obviously cherishes it. It stands there most substantially notwithstanding the question mark of its future. A twelfth-century predecessor 'le Moremyll' belonged to St Albans Abbey and one abbot, John de la Moote, rebuilt it in style in the late fourteenth century, apparently a matter criticized by the brethren as another of his extravagances to the exclusion of other and equally pressing needs of the monastery.

The building dates from *c.* 1760, but much oak, re-used from an older structure, can be seen mixed in with the newer pine members; for instance, the plate laid on the brick ground storey wall is ancient, while some framing at the western end, and near the mill house, looks much older than the main structure. The four roof trusses, similarly, do not seem to belong to the eighteenth-century rebuilding and could be, like the oak roof collars, an adaptation from the previous building. The main part of the construction belongs to its period except that in addition to the usual cross-spanning binder beams picking up the floor joists, the carpenters included two sets of longitudinal beams. Despite this mass of timber, the construction has not avoided failure, because, at some of the beam junctions, stability has had to be guaranteed by inserting supporting posts.

Two plants comprise the machinery. The western plant, driven from an iron breast-wheel, about twelve feet six inches in diameter and eight feet wide, has kept its iron pit-wheel and spur-wheel with wooden teeth, supported on an oak hurst. A smaller iron wheel (ten feet in diameter and approximately six feet wide) drove the eastern plant, all in cast iron, the gear-wheels having wooden teeth, on a line shaft pattern. Here lay a complex system of gearing, of which the present 'stored in' state of the interior hindered inspection. The pit-wheel turned a wallower on the line shaft to which another gear was fixed at the far end, both of these engaging a stone nut and spindle for the two sets of stones.

The spindles carried a band pulley, with another on the line shaft to drive the crown-wheel shaft, and the processing apparatus as well, now removed save for some spouts and ducts. A visit in the thirties showed this interior well filled with these. The gearing design was probably constrained by an earlier arrangement replaced piecemeal — since the eighteenth-century general rebuilding. On the 'stone' floor, four sets of stones remain, two for each plant, three almost complete, while a particularly choice wooden line shaft, driven off the crown-wheel, calls for attention; at least twelve feet long, and eight inches in diameter, it is octagonal and carries wooden cog wheels at each end.

RIVER COLNE

Another well-worked river serving at one time some twenty-eight sites, not counting those upon its tributaries, the Ver, Gade and Chess, yet not one of these older watermills still works. The river rises near Hatfield and the first mill would seem to have been that for Aldenham, listed in Domesday. This site is now unknown though probably it could have been by the weir on the Colne, south of Munden House. At one stage in its history it belonged to a certain Roger Meridene, who in 1220 granted to the Abbey of Westminster all his rights in the mill, the pool and stream, including the obligation of doing repairs. Later in the century, his son or a younger relation, forewent the money rent in lieu of an Easter gift of a pair of white gloves, value one penny, a payment prevailing until the fifteenth century.

Bushey Mill likewise does not survive but at least a pictorial record remains.[15] The façade consisted of three roofs, a flank wall and a miller's residence behind, obviously covering a building of great age. The artist has shown the quaint balustrading of the bridge nearly as high as the mill itself, and also depicts the miller crossing the bridge cautiously as if it might collapse at any moment.

Next comes the town of Watford. The inhabitants, as tenants of St Albans, were required to grind their corn at the Abbey Mill in High Street. The normally submissive relations between town and convent were disturbed in the early fourteenth century when certain men of the town trespassed on the abbot's property. They constructed in the pool elaborate machinery for catching the fish which should have gone to the convent tables. This fishing must have been highly productive for John Wellys in 1431 claimed it as his own property — though unsuccessfully. Watford had a large number of tenants under this abbey who were compelled to use this mill. The latter must have been of a fair size, as repairs for it and Oxhey Mill, carried out by Abbot John of Wheathampstead, cost £41. 3*s*. 6*d*.

The Abbey Mill, after the Dissolution, remained with the Crown for a lengthy period. In Elizabeth's reign a rent of £13 a year from its profits was granted to Sir Christopher Hatton, that favourite of the queen, who probably obtained this annuity by the same dubious means as that by which he acquired Ely Place, near Hatton Garden, from the See of Ely. The same charge continued to be paid by Edward Ferrers and Francis Philip when this firm leased the mill in 1609. Rebuilt in Victorian times, at no gain pictorially or historically to the High Street, a fire in 1928 gutted the structure which for some years stood against the sky, a roofless ruin. These remains were taken down after the last war, a fate to be followed by the neighbouring houses, leaving the immediate vicinity characterless, even for so unprepossessing a place as present-day Watford.

Further down the Colne, near the railway line, stood Rookery Mill. Though not recorded in Domesday, it was probably built before the manor of Oxhey was divided *c*. 1282, when it became the mill of the other moiety. Perhaps it represented the mill also presented to St Albans Abbey by Petronilla de Ameneville and the one given later, in 1276, to a lord of Oxhey Manor.

Oxhey Mill itself has been demolished, but many of the supporting Georgian buildings were kept. The latter, with the gardens, waterways (the last augmented for the nineteenth-century mill) and foliage, together with a sharply modelled site, form a worthy picture. In the thirteenth century, as was then common, the manor became divided, each portion keeping a separate mill. Oxhey Mill remained with Oxhey Manor, suffered damage in 1381 through a quarrel between Edward III's mistress, Alice de Perres, and the Abbot of St Albans over the right of possession, and in 1540, came to George Zowches, from whom it took the name of 'Souches Mill'. He returned it, in 1542, to the Crown and by 1565,

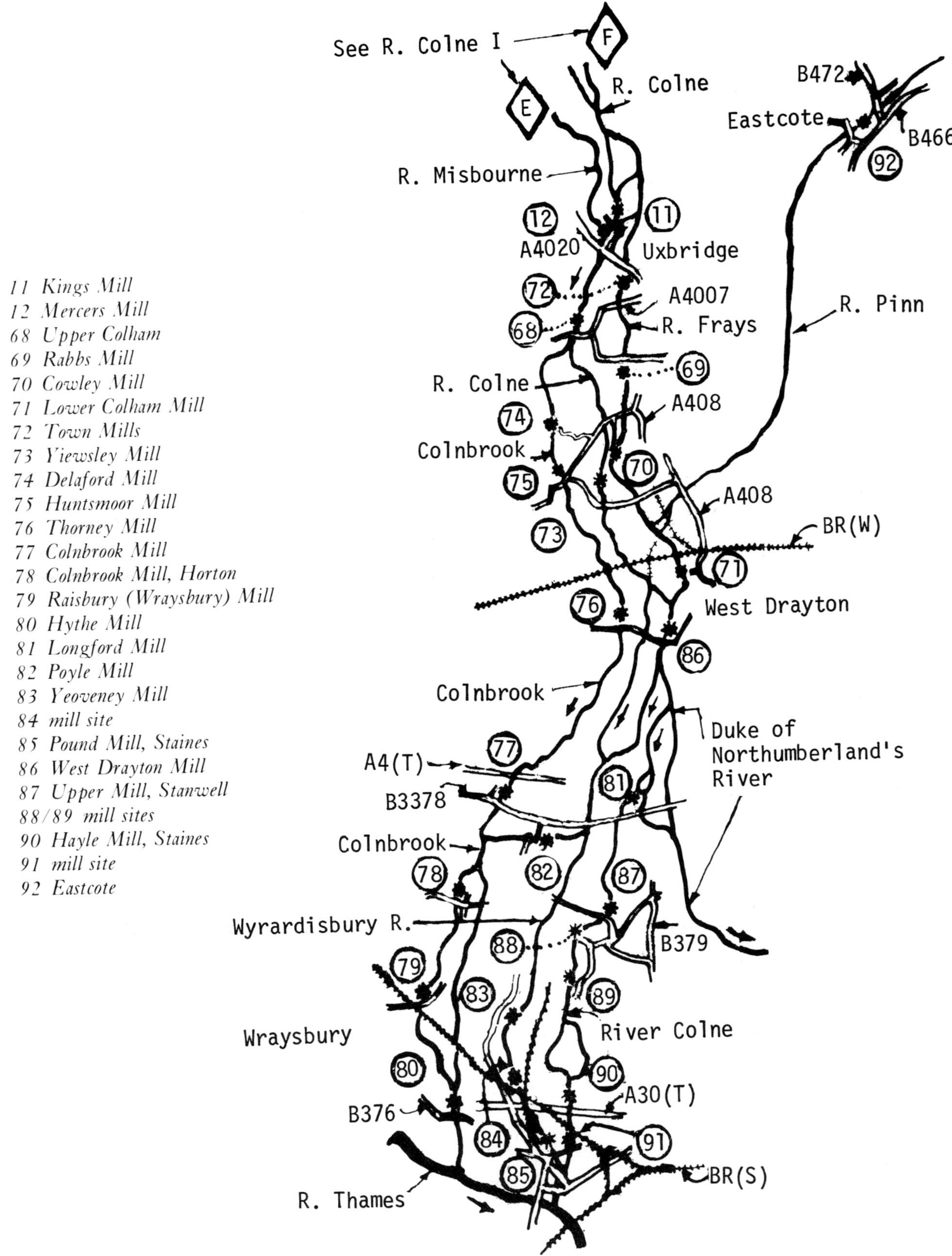

River Colne II

consisting of more than one water-wheel, it had come to bear its present-day name — Hamper Mills.

Rickmansworth possessed one mill at Domesday, but by the middle ages seems to have possessed seven, though identification between those mentioned in the records and a precise site cannot be certain. The first lay on the southern arm of the Colne below Batchworth, and was removed before the First World War, having been used in the early nineteenth century for making cotton goods. It was subsequently converted to paper-making by its purchasers, the Dickinson family. The next, standing at the turn of the road at the bottom of Batchworth Hill, was in the thirties a factory and today seems to have been removed. Mill End Mill, also for paper-making, near that part of the town similarly named, has also gone, though from Domesday it belonged to St Albans Abbey, and was subleased to various tenants. The Colne here breaks into several channels. The channelling, nearly all of which is artificial, suffered alteration when the Grand Union Canal came through, but the mills remained. Firstly those in Harefield, at the locks where the road to West Hyde crosses — modern factory buildings mark the site. Until the late twenties the manufacture was of asbestos goods; previously polishing plate glass; before that it was a copper-mill supplying the navy of Nelson's day with bolts and sheeting. Previously, it had become a paper-mill and earlier still the water-wheel drove pumps, or such like, in a scheme supplying Uxbridge with water.

The Colne divides, below this example, westward to Troy Mills, of which a factory building of *c.* 1860 alone represents a mill shown here in eighteenth-century maps, while the other branch goes to Jacks Mill on the Harefield side of the Valley. The latter, dating from early Victorian times, stood alongside the lock and is now converted into a dwelling. The lade, in 1931, still tumbled through the centre where the wheel originally turned.

The Colne, on its own, flows onward to be joined at Denham by the Misbourne, after which for the next nine miles or so, as far as the Thames, it divides itself into three and sometimes more channels, natural and artificial, in a rather involved pattern and one of great age. Firstly, on the west — the Colnbrook; on the east — the Frays; in the middle, so to speak, the Colne proper, which for the last six miles, produces the Wyrardisbury River. Drainage of this rich valley would seem to have been the motive but equally to get power to the numerous watermills. Domesday registers 25½ below Rickmansworth, though this figure more likely represents wheels rather than buildings.

It would probably be clearer to commence with the Colnbrook, starting from the bridge to the Rockingham Road. The first would be Delaford Mill, now no longer to be seen, while further down stream by a bridge stood Huntsmore Mill till last century.

The fine Georgian miller's house should still exist and the site might mark the mills held in the early fourteenth century by the grandiloquently named Eubolo l'Estrange, lord of Knockin in Shropshire, an ancestor of the Earls of Derby.

The next one, Thorney Mill, had ceased working in the thirties but the present buildings on the site cannot be recognized, or even related to the eye-catching scene that once greeted the traveller. That long shapely mill building could be readily seen from the train. A closer inspection kept the more distant promise, particularly in the colours which really made the picture. Brickwork for the lowest storey, weather-boarded above, carrying a plain tile roof, the normal materials yet the weathering had laid upon them such hues of greys, russets and greens in a way that no conscious selection could achieve, the obvious complement to the subtle, often called featureless, landscape of this wide yet level valley. Inside Thorney Mill, one could see no machinery from its paper-making days, its last trade before becoming a store for the Iver Tube Works, while the framing, with kingpost trusses, in the rood, all in pine and wrought smooth, confirmed its early eighteenth-century date.

Colnbrook, a town which sprang up on the boundary of two parishes in the early middle ages, had a mill in 1274. It belonged in the sixteenth and seventeenth centuries, together with those of Wyrardisbury and Horton, to the Bulstrode family, but flames consumed the last flour-mill some time after 1862. Its site can be observed at the end of Mill Street, with relics like the spillway and wheel-race still to be seen, accompanied by a few of the old houses that survive from this village.

Colnbrook Mill, Horton, now a factory for Meredith & Drew, keeps its pleasant surroundings as did the old mill which in 1930 had been much rebuilt. William de Windsor, lord of Horton, in 1210 leased the mill for four years to Hamon, to his son Henry and their heirs. This grant included the right of grinding for the Horton folk and the manorial household, and this evidence was produced by Hamon at a lawsuit in 1222.

Less than two hundred years afterwards William Blakemore held Horton Mill and in 1390 he diverted the stream close by and upon it built a fulling-mill, evidently adjoining Horton Mill itself. His widow and Thomas Meldreth, her tenant, occupied this new mill for sixteen years after his death, but some disgruntled persons caused an inquisition to be held in 1408. Its decision held that the watermill, then valued at 13*s.* 4*d.*, had been built on the ground belonging to Horton Common. Forthwith Mistress Blakemore and Meldreth were dispossessed, but evidently the mill remained standing. These two victims persevered in trying to regain the property and in 1410 they received a writ of restoration, which certified that the mill had been built after all on Blakemore's own property. Thomas Meldreth still inhabited it in 1428. The tenant following him was John Pury, who, like his predecessor, held the mill property as the manor of Hortonpury.

The Phipp family possessed the mill in the seventeenth century when it was the most considerable paper factory in Buckinghamshire. Edmund Phipp was owner and manager of the works in 1635. That year he offended public opinion by working his mill on a Sunday and, for his enterprise, was summoned and fined by the ecclesiastical court. The neighbourhood held this industry as responsible for much misery in the parish because the casual collection of rags needed in the process often included clothes infected with the plague, thus making dissemination so easy. An outbreak in 1626 claimed thirty-four victims, one in 1636 took fourteen out of a death total of thirty-one, the population being then something around five hundred.

Phipp could do little else than close down Horton Mill. In less troubled times his popularity stood no higher but when, in league with other paper-makers in the county, he petitioned the authorities for a contribution to their relief, they met with a deserved rebuff. The justices of the peace counter-petitioned, not only against a rate relief, but to have the offending mills removed. At the same time, other public complaints arose, alleging that the paper made would not bear ink besides being too expensive. The paper-makers also used their power to bring down the price of rag collecting by stopping their mill; a kind of 'lock out' that apparently succeeded. Herein may lie the origin of Phipp and Co.'s prosperity.

Apparently for a time after making paper, the mill housed a shawl-printing establishment until its destruction in 1839. From this site the river leads to Wraysbury, or Wyrardisbury Mill, a Domesday foundation. In 1547 it was held by the Bulstrode family as tenants of Wyrardisbury Manor, and known as Colnett Mill in 1605 when it had followed the fashion of this valley by making paper. In the *Daily Courant* of 23 June 1722 this advertisement appears: 'Raisbury Mills on Cony stream . . . are to be sold'. A nearby Copper Mill Road indicates a later use. Since the war the buildings have been replaced and all that is of age is the site. The last mill on this branch could be found at Hythe. Its latter-day descendant, a millboard manufactory, remained in the thirties, but only the lower courses of its walls and a cast iron water-wheel survive.

The next arm of the Colne is the Frays River. This one appears to be pre-Domesday in origin, a work some five miles long, cut from the main channel to serve mills at Cowley and Yiewsley, and later harnessed to two others. The mills beside the 'Uxbridge' or 'Woxbridge' from which the town was sometimes called, attracted the attention of early travellers. Leland in 1540 describes both bridges, one over 'the great arme of Colne river' while 'the lesser arm [Frays River] goeth under the other bridge, and each of them serve a great mill. . .'.

Uxbridge, from the eighteenth century, had been noted for its milling trade, due not only to the number of mills, and to the increasing population and relatively easy communications with London particularly after the making of the Grand Union (or Junction as formerly) Canal, but no less to the agricultural productivity of this part of the Thames valley. The first, the rather dreary-looking Town Mills, can be seen at the extremity of the High Street in a decidedly run-down, traffic-bashed end of the town. In the twenties, when thriving, this group of mixed constructions, a late eighteenth-century mill on the original site, and a modern steam-driven version across the canal-head could not be termed captivating. During the last war 'circumstances forced it to close'. Though the waterways remain, forming an attractive feature to the rear, the canal slip, a branch off the old Grand Junction expressly made to bring barges right up to the mill's hoise loft, now looks forlorn as well as being none too salubrious. The old portion now gives shelter to a training centre, whilst amidst all the decay still stands the sixteenth-century miller's house.

The Frays next served Rabbs Mill, now replaced by a factory leaving the mill-race still to be seen, alongside the Cowley Road, a rather forbidding nineteenth-century example which doubtless took its name from the Walter Rabb who, in 1415, founded a chantry in Uxbridge Church. Then comes Cowley mill, closed down at the turn, or thereabouts, of this century, and which belonged to Cowley Hall, a manor held in medieval times by the Abbey of Westminster. Lower walls only indicate the mill, but mill-house and waterways survive still, presenting an attractive picture. Another, once a paper-mill, Lower Colham Mill at the end of a lane of the same name, has been replaced.

To return to the main stream, or earliest course of the Colne, there are two mills on the Denham side of the two Uxbridge bridges, Kings Mill and Mercers Mill, sometime called Mede Mill. The first still remains in the corn trade with all-modern machinery, but the latter shows its origin, through many later additions and conversions. Various sounds of tumbling water pervade this quarter, and despite expectations to the contrary, the array of channels is still a feature. Since 1932 a block of flats has arisen in front of the road bridge and in this conjunction the 'waterscape' may look more amusing than beautiful. Mercers Mill had ceased working by the 1930s but conversion, one side into flats, the other into a car showroom, has at least ensured its survival as a building, though the plant has long been removed. It can hardly be the 'great mill' mentioned by Leland, least of all that one recorded as having been rebuilt in 1389, though another name sometimes given to it, New Mills, likely enough refers to that early rebuilding.

The lord of Colham, who in 1377 received a rent of 'one rose chaplet' for Mede Mill, during the year 1370 stopped a water-course at the mill, thereby flooding the

St Michael's Mill, Kingsbury, St Albans (1932)

Stanwellmoor Mill, Middlesex (1931)

the land of his neighbours. The miller had also committed the same offence. Five years later the Manor Court Rolls records that the 'miller at Mede Mylle' was to be inquired about next court day. Apparently, irresponsibility in controlling the sluices was a characteristic of the millers here. In 1377 'Medemille' was in hopeless disrepair, to be rebuilt in 1389. The neighbouring lord of Colham rented it in 1396 for a farthing, or a red rose. Names of millers can be given: Roger Morecok in 1401; Richard Clifford in 1440; and in the following year William Dyer, who also owned several portions of land in Denham. William Stanysford was the lessee in 1450 and John Fabyan the miller in 1501.[16]

Two watermills known together as the Abbot's Mills were situated on another branch of the Colne until their destruction by fire in the mid-eighteenth century. Today Kings Mill stands by their site. The western one was a fulling establishment, while the other alongside was known as the 'Corne mill of Denham attached to Colne', later as New Mill. Apparently it was the chief one in the parish. Roger the dyer of Uxbridge held both in 1303. In 1375 the convent of Westminster granted the corn-mill, then known as Old Mill, and the dyeing works situated upon the 'dyeing mill eyot', to Roger Morcok, who must be the same person as the holder of Mede Mill in 1401.

In 1388, when the fulling-mill fetched £8. 0*s*. 0*d*. in rent yearly, the corn-mill was so old and rickety that no tenant would take it. Perhaps the old structure was rebuilt immediately, because afterwards both mills were always let together by the abbot. John Slyng was in possession here in 1388 and John Dundle in 1396, while William Clifford (probably a relation of the Richard Clifford at Mede Mill in 1440) took the mill in 1455, when his assistant was William Dyer, junior. Was he the son of that William Dyer, lessee of Medemille, after the above Richard Clifford? William Clifford still worked the mill in 1478 and by 1501 John Fabyan had taken his place, who concurrently held the same position at Medemille.

No substantial remains tell of Upper Colham Mill near the canal by Rockingham Road, with only a little more of another, that just off the Green at West Drayton, a large paper-making concern shown in 1754.[17] The ruins of the latter today (and appearances have not changed much for forty years) belong to a building much later in date. Below this site, from the Colne, there issues forth another branch, the Wyrardisbury River, and also the Duke of Northumberland's river which a few miles eastward supplements the Crane.

Wyrardisbury first drove Yiewsley Watermill, now removed, and then it came to a watermill at Longford on the north side of the old Bath Road, which survived until at least 1754.[18] Earlier still, it belonged to the priory of Harmondsworth. This alien cell was confiscated in 1340 when it possessed a corn-mill, value 18*s*., and a malt-mill, value 8*s*., probably under one roof. The tail from this last and vanished mill led to the pool of another, Poyle Mill. Today it is only a factory occupying an old site, that of Richard West's paper-mill of 1636 and earlier still, of a mill mentioned in an inquisition of 1424. Before reaching the Thames, the Wyrardisbury drove more watermills. First came Yeoveney Mill, now without known site, though probably near the manor house of that name. It seems to have been removed by 1376. A 'New Mill' is recorded on an unnamed site, possibly by the railway crossing, while the last mill on the river stood alongside the closed railway station. The lade here can be seen and a rather dreary sham half-timber structure housing the pen-stocks. Pound Mill, a name denoting the original use of the area now surrounding it, was demolished in 1916.

To return to Colne itself, the next to be mentioned is Upper Mill, Stanwell. The buildings of this example, in 1931 a dyeworks, hardly compelled a visit, but most impressive then, and no doubt for some years following, was the large undershot water-wheel, something like twenty feet in diameter and six feet wide, with plashers or paddles eighteen inches deep, when revolving in its repetitive, seemingly effortless, fashion, giving forth so little sound for so large a machine, with the discarded water gliding down the tail race to finish in the pool on the other side of the roadway in a tumult of foam and wavelets. This sight is no longer to be enjoyed, and there has been some alteration to the mill itself. This example may descend from an illegal mill erected in 1385 alongside the bridge over the Colne, described then only as a common footway for passengers. The builder, John Donet of Stanwell Moor, was presented before court for this flagrant breach of the law but, though the verdict ordered him to remove it, the mill was evidently suffered to remain.[19] A certain 'Ralph atte Mille', mentioned in 1398, could possibly be associated with this mill as could also a paper-maker of the seventeenth century, Eustace Burnley.

'New Mill' mentioned previously, might have stood on the main stream by Hammond Farm, where the Bonehead Ditch joined it, but today the bypass seems to have changed the topography so much that checking on the ground becomes unrewarding. Then comes the second mill at Staines, the last on the stream, a shell only, and sometimes called Hale Mill. It has for many years formed part of a complex factory group manufacturing linoleum but recently closed down. The water-wheel disappeared long ago and the building keeps much of its eighteenth-century walls and roof, although the inside has been gutted, and it is, by all indications, unlikely to survive despite being the last of some six mills recorded in Domesday for Staines.

NOTES AND REFERENCES

1 Bryant, op. cit.
2 Bryant, op. cit.

3 Lathbury, op. cit. (Chap. 5, footnote 28)
4 Since this was written the mill has been severely damaged by fire
5 Ellis, W., *The Modern Husbandman*, 8 vols., London, 1750
6 Walsingham, T., *Historia Anglicana*, vol. III, Rolls Series, Longmans Green, 1863, p. 263; cf. account in Victoria County History *Hertfordshire*, vol. IV, pp. 198–207
7 For a view see p. 22, A. W. Ball, *Watford Pictorial History*, Watford Corporation, published 1972
8 Walsingham, op. cit.
9 The following paragraphs are an interpretation. See Victoria County History *Hertfordshire*
10 Andrew and Wren, *Map of St Albans*, 1766
11 *The Ambulator . . . a stranger's companion on a tour round London*, J. Bew, London, 1774
12 For illustration before destruction, see *Hertfordshire Advertiser*, 5.12.1958
13 Walsingham, op. cit.
14 Chauncy, op. cit.
15 Chauncy, op. cit. A slightly earlier view (British Museum Maps, K XV, 72.1) shows a thatched mill, one storey and a half high, with, characteristically, a mill-stone leaning against a wall
16 Lathbury, op. cit.
17 Rocque, *Map of Middlesex*
18 Rocque, *Map of Middlesex*
19 See p. 53 for a similar case not far away, e.g. Colnbrook Mill, Horton

Staines Mill, Middlesex (1932)

Chapter 16
THE LEA AND TRIBUTARIES

RIVER MIMRAM (or MARAN)

THIS chapter deals with watermills and their sites in the Lea basin, which is well supplied with tributaries. A start will be made with the Mimram, followed by each tributary in turn, and then the main river.

King's Walden Mill, cited in 1313, has no representative today but the nearby village, St Pauls Walden, had an example at the hamlet of Whitwell. Burnt down in about 1900 the somewhat ramshackle replacement has also given way to sheds, but the iron water-wheel can still be seen in situ. This manor and mill and another one further downstream were mortgaged for ten years in the early fourteenth century by that incompetent abbot of St Albans, Hugh, to William Leggatt, this man being probably a relation of Hugh Leggat, a learned monk and a historian of the abbey. The last tenant of the abbey was one Thomas Ventor who, in 1538, also held two meadows and the multure due to the mill. A long wall and a sluice at the crossing of the Mimram by the Welwyn Road were once to be found as the sole indications of Pann Mill, while nothing remains of the Knebworth Mill mentioned in 1613, which probably stood near the lake in Hoo Park.

Kimpton Mill follows next, situated alongside the roadway. Here, in 1931, the machinery and an undershot water-wheel, all of wood and a rather rare example, could be seen. The structure of the mill itself was not large. Seventeenth-century brick groundwork carried a weather-boarded storey above which, in its timber framing and like the miller's house next door, would be much older. How long this state survived is not known, but since the last war, the structure has been gutted and converted into a residence and, though the original building in the main still stands, something of the former spontaneity, then readily evident, has gone.

A white dwelling house incorporating Codicote Mill beckons across the meadows, though the more exciting approach from the village, is to be preferred. The mill itself was rebuilt and converted before the First World War and was originally about three centuries old. Today it can only be recognized by the fact that it stands over the stream. A predecessor was a victim of the financial mismanagement of Abbot Hugh mentioned previously, though his successor apparently repaired the mill, which later figured in the revolt of 1381. While never discovered, the incendiary who set fire to Codicote Mill was thought at the time to be a relative of 'John Biker', one of St Alban's disaffected tenants. But the miller by his promptitude saved it from destruction. The next abbot, John de la Morte had, it seemed, to order its rebuilding in 1397.

Before the Mimram reached Welwyn, it drove a mill at Ayot St Lawrence, recorded as out of repair in 1354 and completely ruinous in 1375, which may be why its site is doubtful, but probably was near Pulmore Water. Just north of Welwyn village can be seen the early nineteenth-century shell of Fulling Mill. It functions as store and workshop for a boat-building business. By 1931 it had not been used for many years but the plant then remained including an overshot water-wheel.

Welwyn Mill, which stood until the last century in a lane east of the town, was mentioned in 1469 when a certain Thomas Dean of Ayot Montfichet (Ayot St Peters) 'burgled' the mill to the rector's damage, the latter being lord of the manor. Two years later Thomas Payne apparently raised the level of the weir at this mill and flooded the King's highway which would be the present street; an occurrence not unrepeated since.

Digswell Mill stood below the railway viaduct south of Welwyn Station. The Georgian mill had been converted into a dwelling well before the First World War and between it and the mill-house was sited the water-wheel. Peter Valognes held the mill at Domesday and it was mentioned again in 1243, while another mill stood in the parish as late as 1786 but with an uncertain site, probably at the weir in Digswell Park.

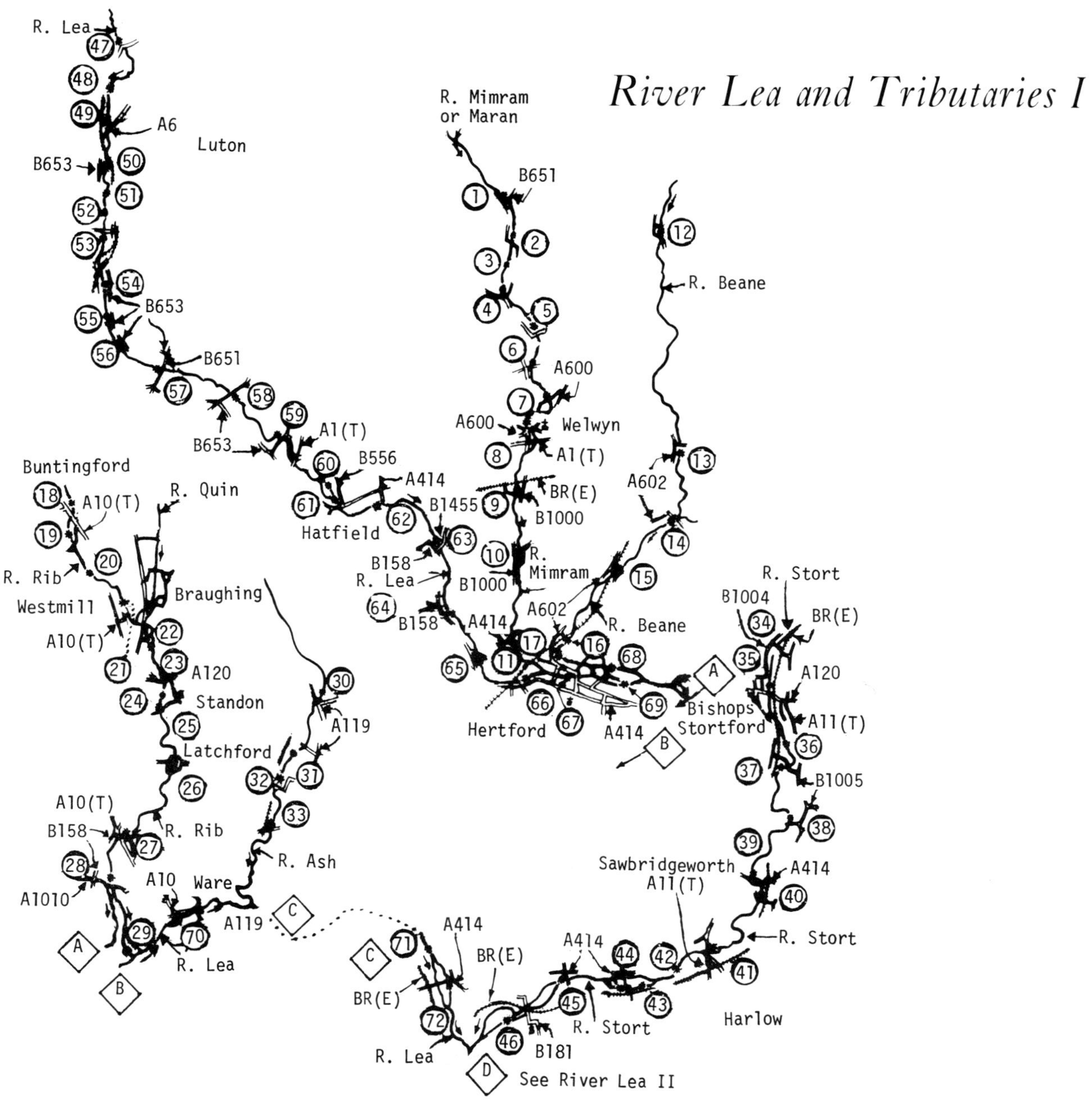

1 *Whitwell Mill*
2 *Pann Mill*
3 *Knebworth Mill*
4 *Kimpton Mill*
5 *Codicote Mill*
6 *Ayot St Lawrence Mill*
7 *Welwyn (Fulling) Mill*
8 *Welwyn Mill*
9 *Digswell Mill*
10 *Tewin Mill*
11 *Hertingfordbury Mill*
12 *Walkern Mill*
13 *Watton at Stone Mill*
14 *Sacombe Mill*
15 *Bulls Mill*
16 *Molewood Mill*
17 *Sele Mill*
18 *mill site*
19 *Aspenden*
20 *Hull Mill*
21 *Hamels Mill*
22 *Gatebury Mill*
23/24 *mill sites at Standon*
25 *Old Paper Mill*
26 *Latchford Mill*
27 *Wades Mill*
28 *West Mill*
29 *Ware Park Mill*
30 *Hadham Mill*
31 *mill site*
32 *Widford Mill*
33 *Mardock Mill*
34 *Parsonage Mill*
35 *Town Mill*
36 *South Mill*
37 *Thorley Mill*
38 *Little Hallingbury Mill*
39 *Sawbridgeworth Mill*
40 *Sheering Mill*
41 *Harlow Mill*
42 *Latton Mill*
43 *Burnt Mill*
44 *Parndon Mill*
45 *Hunsdon Mill*
46 *Roydon Mill*
47 *Leagrave Mill*
48 *Limbury Mill*
49 *North Mill*
50 *Bury Mill*
51 *Brache Mill*
52 *Stapleford Mill*
53 *New Mill*
54 *Hyde Mill*
55 *Pickford Mill*
56 *Batsford Mill*
57 *Wheathampstead Mill*
58 *Sandridge Mill*
59 *Lemsford Mill*
60 *Paper Mill*
61 *Mill End Mill*
62 *Cecil Mill*
63 *Essendon Mill*
64 *Waterhall Mill*
65 *Bayfordbury Mill*
66 *Horn Mill*
67 *Town Mill*
68 *'Flower' Mill*
69 *Dicker Mill*
70 *Ware Mill*
71 *Amwell Mill*
72 *Stansted Abbots Mill*

Kimpton Mill, Hertfordshire (1931)

Lower Mill, Codicote (1931)

Welwyn Mill (1931)

The Mimram flows towards Tewin Mill. Manor and mill were bestowed upon St Thomas's Hospital by Geoffrey de Tewin in the fourteenth century helping to support that establishment until the Dissolution. This mill was a double one in 1368 and then called, strangely, 'Le Solo'. Together with the manor it was bestowed by the Crown in 1544 on John Cock and, as the last Tewin Mill, was demolished in 1911.

The river then flows through Panshanger Park to the mill at Hertingfordbury village. It stands right in the traffic line through the village street which, with the usual press of modern vehicles, prevents an easy view of it. It was working until recently, as the display of odd sheds and apparatus attached to it indicates, but it is now closed down. Apparently some of its machinery is still there. The building, dating from about 1885 and built in the Suffolk whitish brick common to many mills in the eastern counties, does not look all that elegant. Of Domesday origin, in 1247 it was leased to Henry Neketon by the manorial lord and some time afterwards was given by Peter de Maule and his wife, Christine de Valognes, holders of Hertingfordbury Manor to the hospital of St Mary Magdalene in Hertford, founded by a member of that family for the Trinitarian friars. These two persons again seem to have rented the mill back from the hospital, not unusual in bequests of those days, and in 1279 Christine paid £1 to the master of the hospital, a rent still being paid in 1354 and again in 1383 when it had risen to £2. 3*s*. 4*d*. This particular foundation, unlike most medieval hospitals, did not survive to the Dissolution because the Crown had taken it over in the fifteenth century.

RIVER BEANE

Though the mill of Weston, whose miller had been fined in 1201 for filling in part of the pond thereby inundating the highway, stood on this river, the first mill to be seen today lies south of Walkern village. The structure survived after the business had closed following the last war. A relatively new building, *c* 1870, its lineage stretches back to Domesday and to its gift in the early twelfth century to St John's Abbey, Colchester, by Hammode de Clare. The pool has contracted to a stream's width but the pleasant tumbling bay in a convex brick wall, probably as old as the mill itself, survives as an unobtrusive, but not less delightful, example of water landscape architecture.

Fortunately, the owner, Dr Bugge, intends to keep the building, and eventually adapt it to a suitable use, which objective, to judge by the successful renovation of the adjoining one-time miller's house, is heartening. That reinforced concrete monstrosity which almost hides the watermill from the south is scheduled for removal. Its existence from the early years of this century proved how good were the trade prospects for this mill at that time.

The watermill in Watton at Stone, which with its late nineteenth-century shape, stood behind the village street, seems to have gone. Until Hertford, the Beane passes sites only: those of Saccombe Mill at Mill End; Bengeo's mill at Bulls Mill, Stapleford. The latter may be identified with a drawing of *c.* 1794 which shows a three-storeyed complex of weather-boarded walls and tile roofs (demolished 1869). Molewood Mill, near the waterworks, comes next and part of the last mill here survives, much altered when this particular statutory body took it over in 1868, while an arch keystone dated 1801 marks the race. The modern roller-mill at Sele Mill, prominent from the main road junction, has special interest in its history for here must have stood the first paper-mill erected in England — the one referred to in one of De Worde's printings. The first indication dates from 1496, while in 1498 there is reason to believe it was visited by Henry VII when during the same year 'a rewarde' of 16*s*. 8d. was given to the 'paper mylne' and the next year saw a further gift of 6*s*. 8*d*. The owner of this pioneer venture was John Tate whose father was lord mayor of London. It was working when Tate died in 1507. He instructed his executors to sell the mill, complete with appurtenances and thus the manufacture ceased, or at least records remain silent. By the eighteenth century there was on the site a small building with an external water-wheel,[1] to be replaced *c.* 1820 by the predecessor of the present one.

The River Ash has insufficient importance to call for a separate heading, because, of the four watermills once dependent on this stream, only one survived as late as the thirties, that at Hadham Farm. Some lower walls of Mardock Mill could be seen below the equally defunct railway station of the same name, while the two other sites lay at Widford Mill below Blakesware, and by Mill Mead in Widford.

RIVER RIB

At the eastern extremity of the gardens behind Buntingford's main street the Rib tumbles over a weir. Perhaps the mill given to the nuns of Holywell, Shoreditch, in 1239 stood at this spot. Equally no relics remain of Aspenden Water Mill, once to be seen at a short distance from the now defunct Buntingford railway station. West Mill, a pleasant village, belies its name for of its original four mills the site of only one remains, situated in Coles Park where the name 'Mill Croft' recalls the old Hull mill that disappeared last century. Amidst the relative fastness of a wood close to the bridge carrying the Royston road over the Rib, the site could be recognized from a weir, with a few yards away a dried-up overflow channel.

Hammels Mill, Braughing, had been derelict some years in 1932, as loose slates, broken windows and unhinged doors then proclaimed, but its site, below where the Quin joins the Rib, seems all there is now. The most important of Braughing's mills, was that at Gatesbury, which was destroyed in 1906. Queen

Paper Mill, Sandon (1932)

Payne's Mill, Ware (1932)

Walkern Mill, Hertfordshire (1931)

Matilda presented the predecessor to the hospital of St Katherine by the Tower, to be exchanged by Stephen a few years later for other land and then returned by Henry de Durneayx and Theobald de Braughing to the hospital. It remained with the brethren until the Dissolution but by 1656 it had become a fulling-mill and, in that year, it was sold to Robert Dicer of London.

Four mills appear to have met the needs of Standon, the next parish, which until the Black Death, was a considerable market town. Until the late eighteenth century three mills stood close to the village itself and all seem to have been at one time directly held by members of noble families: one by Joan, widow of William de Clare, two by Elizabeth de Burgh, and also the fourth by the Knights Hospitallers. In 1398, when decay had overtaken the little town, only six customary tenants could be found to support these four mills. The uppermost mill on the river has a late Victorian building, carrying the date 1901 on it, yet the situation was once much more picturesque. This represents the chief manorial mill and perhaps that belonging to Joan de Clare in 1307. Downstream, and situated by the gardens of the houses in the market place, stood a second example called 'Lynch Mylle' which was a fulling-mill in 1362, but of this nothing remains. To the south of the village is a third called the Old Paper Mill.

This was only a nineteenth-century single-storeyed structure, in fact a rather insubstantial sort of building, with the wheel-house alongside. The adjoining mill-house, much older, with the accompanying ford and waterways made an attractive picture in 1930, and to a considerable extent still does. There remains, nearby, one of the many fords that the river Rib provides as it goes through this part of Hertfordshire. In those days the water-wheel had started to collapse and today little more than a few timber pieces survive. Its predecessor in the middle ages belonged to the Knights Hospitallers.

The Rib then passes through the hamlet of Latchford where in 1362 Elizabeth de Bere held 'Lotiforde Mill', but by 1371 it was called 'Cuttymelle' and belonged to the manor of Plashes. Today signs of this example are nil.

Next on the river was Wades Mill, Thundridge, belonging to Odo of Bayeux at the time of Domesday and called 'Walysmill' at the end of the sixteenth century. Until a fire some time ago its Victorian bulk stood below the Roman road at a spot where one of the first of the turnpikes was placed. A local proverb once said 'Ware and Wademill are worth all London', and the puritan susceptibilities of old Fuller the historian were so ruffled at this as to provoke his reply 'that it were a masterpiece of vulgar wit'. Of the structure it seems that only a doorway and a 'lean-to' portion on the right-hand side can be recognized as belonging to this particular example. The rest of the site now seems little more than a scrap dump, while the fine chestnut trees, like others which seem to belong to mills in this valley, the old ford that once passed over the tail water, and the tail arches, have been obliterated.[2]

West or Payne's Mill, on the Rib amid the few houses near the crossroads, now ranks as a show place, being rescued in 1962 for a dwelling and quite sympathetically altered and renovated. The remains of the wheel rims and the spokes on the axle have been retained as a feature in the entrance. In the 1930s its four sets of stones still ground, one for flour and two for coarse feed, and most of the gears were of wood, though obviously of a later date than the seventeenth-century building.

Below West Mill starts the lade which fed the mill belonging, from early days, to Ware Park. It had become a fulling-mill by 1360 and afterwards came to the Crown. Henry VIII leased it for forty years to an inhabitant of the town, Thomas Leonard. His lease must have been extended, for in 1587 he disposed of it to Thomas Fanshawe of Derby, a successful lawyer and lord of Ware Manor at this date. One of his sons, Henry, succeeded to this property which, under his guidance, became noted for its flowers, fruits and herbs. He was brother to Sir Richard Fanshawe, the poet and author, who must have known the vicinity well and have often passed by the mill. Ware Park Mill eventually reverted to its original owners, for late in 1646 it still belonged to the Leonard family. The site of its successor may still be found between the hill and the canal, distant halfway between Ware and Hertford.

RIVER STORT

This important tributary rises near Clavering and joins the Lea near Rye House, but of the many old water-mills whose wheels it turned, only the shell of one remains. Mills, once at Hertisham, Stansted Mountfitchet, Bentfield and Elsenham do not survive. Of the latter there was only a mill-house in 1876. All three mills which were at Bishop's Stortford had been removed before 1900: Parsonage Mill representing the Parsons Mill of 1651, then belonging to the Rectory Manor, stood on the north of the parish close to the main railway line; a weir and sluice on the south side of Bridge Street near the castle ground make up the sole remains of Town Mill; while Southmill Road marks one that occupied a site where the bridge spans the canal.[3]

The Stort next supplied Thorley Mill and a note from 1932 can be quoted: 'a stern Victorian building, but when seen across the stretch of the canal, it harmonizes with its surroundings. A piece of glazing has broken away from one of its windows and through it can be seen the water-wheel, an overshot specimen twelve feet in diameter and ten feet wide which in its revolving, verily rocks both wall and window frame'. Today, no longer a mill, an external pseudo-Mediterranean decor hides its origin, while a series of

Wades Mill, Thundridge (1929)

Essendonbury Mill (1929)

Old Mill, Harlow (1932)

Hamels Mill, Braughing (1931)

Thorley Mill (1932)

flats and dwellings now occupy the space where gears, shaft and stones once turned.

Further on the river, Little Hallingbury Mill no longer works but stands very defiantly as an attractive picture at the end of the lane leading to the marina now established on the river. Of long history, mentioned in Domesday and with the usual manorial associations, it belonged in 1570 to Anne, wife of Sir William Parr, Marquis of Northampton and brother of Henry VIII's last queen. Later, in 1720, a silk-throwing or twisting establishment was set up at Little Hallingbury Mill. A company of speculators, having purchased the patent rights of the new machinery from its inventor William Aldersay, were able to monopolize this new form of silk production. A lengthy prosperity enabled employment to be given to many women and girls, mostly from Bishop's Stortford but, after 1860, owing to free trade, falling custom caused it to be closed down. It was rebuilt as a corn-mill in 1874, and this is the building there today which, thanks to its owner, is being renovated piece by piece. It would seem that the previous mill stood adjacent to the present one riverwards, and presumably was removed when the present one together with a new lade had been completed. Internally the well-carpentered pine frame can be seen, placed on the customary brick base, and in fact, this prevalence of wrought pine provides one of the features to be appreciated. Weather-board covered it may be, but vertically, 'baltic' fashion, and the evidence of decay that has taken place at the plank join suggests the reason why horizontal boarding eventually took its place. With regard to the plant, a cast iron water-wheel, shaft and gearing, still survive though some of the stone nuts have been removed, but the arrangement illustrates the common practice when, technically, stone milling had reached finality as far as the country mills were concerned.

The next job for the Stort lay in driving the mill at Sawbridgeworth, but this is no longer true today. Only the mill-race and spillway remain, though the mill-house survived. The adjoining mill yard, with several 18th-century storehouses (now made into dwellings) still attracts. There was quite an array of lucams or hoist lofts, jettied, it would appear, for several feet beyond the wall faces below. Further downstream, Sheeringmill Lane is met and here, until the First World War, could be seen the small example suggested in the name of the lane. An earlier one would have been the mill of Cuwick (i.e. Quickbury, a farmstead on the Essex side of the Stort) which belonged to Bermondsey Priory.[4]

Harlow Mill comes next, once held by the Abbot of Bury St Edmunds. It remained between wars and probably represented alterations made in 1793 when the water, taken from the canal as a lade, was also utilized to bring the grain barges right up to the mill. Later, a miller put a large porch on stilts over this dock, so that unloading went direct from barges into the bin floor. Some time around 1933 the miller's house was formed into the present restaurant and inside can be seen a painting of the mill in 1861 which displays little difference from the drawing of 1932. The mill itself has been pulled down. Mainly built in the early nineteenth century, one can still notice a sixteenth-century post and spur, and also remains of the clasp-arm water-wheel with a piece of the vertical shaft tumbled over it. The Stort next passes the lock where Latton Mill once stood, and this may be the site of Gilston watermill, otherwise unknown.

Next on the Stort comes Burnt Mill, Netteswell, so named because fires consumed it twice in the eighteenth century. Also modern is Little Parndon Mill, while further downstream must have stood the watermill and fulling-mill valued at £2 in 1301, which were probably under the same roof and belonged to Eastwick. At the death of Laurence de Tany, the lord of Eastwick manor, his widow Margaret received, among other property, a third part of the rent from a watermill leased by Humphrey de Walden; but as it had fallen down, this part of the gift was somewhat worthless. As mention is made of it in 1607, and in 1641, when Sir John Gore was the owner, it was evidently rebuilt. The site is uncertain.

The last two to be found on the Stort belonged to Hunsdon and Roydon. The first, mentioned in 1297, stood further down close to the canal lock at Mead Lodge. It was called 'Wades' or 'Wardes' Mill in the first part of the sixteenth century and was held, together with Eastwick Mill, by the manor in 1607. Roydon Mill, near the lock, is an industrial building of about 1830, but apparently no longer works.

RIVER LEA

Next to be treated is the main river, rising in the end of the Chilterns which, from Luton onwards, drove many mills. This town's phenomenal, if now only too typical, expansion has obliterated almost all indications that it had a long past. Only a noble church recalls this. Hardly a vestige can be pointed out of its ancient farms, inns and cottages and also only one of its train of watermills. Undoubtedly these watermills brought trade, indirectly perhaps, to the small town, laying thereby a base for the place's subsequent prosperity. In view of the smallness of the Lea in its upper reaches, it is surprising that within seven miles of its source it allowed the early settlers to impound sufficient water to turn so many mills. In this good corn-growing land, no watermills existed nearer than those in the Ouzel valley to the north or on the Hiz to the east. Hence it can be seen how the Lea watermills met a need.

The Lea filled some ponds by Leagrave Marsh Farm, and the first of the mills stood there, though the waterpower available at the source of the Lea could have been barely sufficient to drive any wheel other than a small undershot, or maybe a primitive Norse type of mill. Leagrave Mill may be identified with the one mentioned in Domesday which was later given by Henry III to a servant of his, Richard de Gray, as a form of pension. The last mention of it was in 1224; later on it must have fallen into decay and disappeared.

A similar fate befell another watermill, also recorded in Domesday, which stood opposite Limbury Manor House. It was functioning in 1247, because at that date the miller was stabbed to death by a robber. By 1368 it was 'worth nothing for want of repair', and in 1408 still remained valueless for 'it was thrown to the ground' and was never again rebuilt. North Mill was Luton's chief corn-grinding establishment in its day; it belonged directly to the manor and flourished until the arrival of the Midland Railway, before which, both building and pool had to give way. Its site can be found near the weir in Mill Street in the town.

Considerable imagination is required to visualize the Lea flowing through Luton with meadows on either side in the place of the present brick and concrete walling. Yet such surroundings once existed, and among the meadows below St Mary's Church stood a watermill named variously Church, Bury or the Abbot's Mill. Unlike all the others, which were held from the Crown, Church Mill belonged to St Albans Abbey. The earliest mention of it is in the Domesday survey, and in 1194 it is referred to as belonging to the abbot of St Albans. Abbot Richard Wallingford had it repaired in 1333, probably after flood damage caused by a severe storm in July 1332, yet at the Dissolution it was derelict and was never rebuilt. During its period of usefulness, however, the abbot vigorously enforced the church's right, not only to tithe corn, but also of compelling the tenants of the abbey lands to bring all their corn for grinding to the church mill and to surrender their domestic hand-mills.

Further south, by Brache Road, stood a mill of the same name until the latter half of last century, when it was pulled down. It was known as the Brache Mill as early as the twelfth century. During a terrible flood in 1795 the owner and miller, John Brown, rode to the mill to help his nephew, who had gone previously to adjust the waterhatch gates, and let the water away to prevent the flooding of the mill. In wading through the waterway to the mill door, his horse slipped and both mount and rider were drowned.

Stapleford Mill lay in Luton Hoo Park and was either at the weir midway in the lake, or at the lakehead, by the lodges. The original of this mill was

Hyde Mill, Luton (1952)

Batsford Mill, Harpenden (1929)

mentioned, like the others, in Domesday and again in the year 1247. It finished its career when the Marquess of Bute had the park landscape laid out and the extensive sheet of water was formed by the celebrated 'Capability' Brown in the late eighteenth century.

One other mill was set up at Luton later than the Domesday survey, probably in the fifteenth century. In 1461 it was referred to as the New Mill, and Abbot John of Wheathampstead purchased a manor and seventeen acres of land there to receive and store his tithes from that part of London. It survived until around the turn of the century, and those who remembered its appearance considered it to be the original building. It gave a name to the present hamlet of New Mill End.

The last of the mills within the town boundary, and the sole survivor, Hyde Mill forms a pleasant group, mainly in the local purple-hued bricks, much improved by weathering in the last forty years. In 1929 it was not working, and then the plant, including a wooden water-wheel, which dated like the structure from the mid-nineteenth century, had been kept *in situ*, though corn milling started again later.[5] The Lea flows from Hyde Mill into Hertfordshire, first to Pickford Mill, and then to Batsford Mill. The first has vanished, and the second, a nineteenth-century building in the pleasant local brick, ceased its trade long ago and in the last forty years has gathered round it a collection of sheds. Next comes Wheathampstead Mill, a long range of buildings. A severe fire in 1900 resulted in certain rebuilding, and it stopped grinding long ago, to be used today as a store and premises for making up products processed elsewhere. Externally there seems little change from its appearance forty years ago.

Internally one can see that the plant has been removed and the original timber and brick structure much added to — for instance, some early weather-boarding can be seen inside a later extension, where the position of the pit-wheel is evident, and also the brick walls underpinning the original oak framing, much of which the builder kept. Two of the original seventeenth-century dormer windows survive at the rear. Walking around at the rear, upstream side, shows how pleasant the waterways have remained. All the same, the prospect can hardly be overlooked that when the time comes for heavy repairs to be made, the business may consider the expenditure not worth it, while the mill's siting, in the narrow roadway, provides a tempting morsel for road widening.

Sandridge had a mill and Chauncy[6] depicts it in his map in Brocket Park, a site which could be identified with the pump-house and weir there. Also the mill (with the manor) leased by Robert Albyn of Hemel Hempstead in 1331 from St Albans Abbey, and which at the end of the fourteenth century was rebuilt by the abbey, may have stood there. Lemsford Mill, a large structure within the new town of Welwyn, represents one of the four mills credited in Domesday to Hatfield, and recorded as a double mill in 1277. Like so many within the economic pull of the capital, it made paper in the early nineteenth century. When that closed down, the present edifice as a flour-mill, replaced it. The flour-milling trade has been replaced by that of light engineering, but this has not spoiled the still-sightly conjunction of bubbling tail water, mill building, miller's house and a cottage. Built about 1880 or so, with the lower storeys in gault brick and the upper ones in weather-boarding, the effect enhances the appearance of this hamlet.

Further down, in the vicinity of Mill End Green, where road junctions interweave in a bewildering manner, two sets of watermills could once be found, the northernmost one being a paper-mill described as disused in 1873 and since that time removed. The second, properly called Mill End Mill, with its adjoining cottage, still survives. It is one of the pleasant features of this much altered neighbourhood, though now a residence, after being a store in the thirties, the corn.grinding activities having ceased shortly before. Externally alterations have been minor though the lucam has gone.

Most of the interior remains and is interesting because of the evolution of the building and the immense scantling of the framework members of the hurst. Had these latter been in oak, such a size would be expected, but here the carpenters used pine, seemingly about 1820. The water-wheel can also be seen in a cast iron frame, with wooden arms carrying the steel buckets, the bottom ones having been removed to allow the stream to flow through unimpeded. The plant has lost the stones and stone nuts; but still to be seen (somewhat cluttered, as the old place provides a useful store and work place for the owner's household) are a cast iron pit-wheel with wooden cogs, the wallower in cast iron throughout, and the cast iron crown-wheel with wooden cogs.

Coming to the building itself, the lower storey, mill-race and channel probably can be dated to 1762, a figure cut in on one of the bricks. The binders carrying the stone floor are of the same age but odd trusses and plates remain from, it would seem, a fifteenth-century predecessor, then a storey and a half in height. The next stage, either *c.* 1825 when the plant was renewed, or earlier when the brickwork was erected, was the raising of the roof to introduce an attic and bin floor utilizing the old plate and much other timber which resulted in the two-storey mill now to be seen. Its owner has looked after it well, and long may it so remain.

Robert Cecil, Earl of Salisbury, when he bought, and first resided at, Hatfield House, started a factory here in 1607, in order to employ the idle paupers of the parish. To set up this project where all the subsidiary process of wool cloth-making should be carried on, he entered into an agreement with a merchant named Walter Morrall. The earl bound himself to provide the premises of which the Mill End Green mill was to be used for the fulling process, and also to disburse £100

Mill Green Mill, Hatfield (1931)

Ponders End Mill, Enfield (1931)

Broxbourne Mill (1929)

Wheathampstead Mill (1929)

Lemsford Mill (1929)

Cecil Mill, Hatfield (1931)

yearly for his expenses to Morrall, who in his turn was to employ fifty persons of the parish. The venture flourished until Cecil's death in 1612, one of the causes of its failure being, as Morrall complained, that parents took away their children who had been apprenticed, in order to earn ready money working in the fields. It however closed down by the next reign.

The river passes through the Home Park to the west of Hatfield's mill — Cecil Mill, still as uncompromisingly unlovely as it seemed in the thirties, though the riverside view and its sylvan accompaniment looks as good as ever. Though of Domesday recording, it dates from 1870, apparently as a saw mill for the estate. Essendon Mill, the next, stopped grinding corn some time between the wars but now forms an attractive residence, the latter using the miller's house and an adjacent part of the mill. Some of the plant remains but not the wheel, as the pool and tail have been filled in. As a group under one roof it contains the miller's house of the sixteenth century; the mill proper, which is a mixture of seventeenth- and eighteenth-century brickwork; and an addition of *c* 1850, on the south end.

The Lea flows onward to the Thames and leads to only two mills that remain in anything like their nineteenth-century shape, but it passes many sites whose historical content is no longer, alas, matched with buildings of equal degree. Next came Water Hall, where stood Hertingfordbury's second mill, while the Bayford Mills mentioned in 1226 have gone — perhaps they stood on the Lea by Bayfordbury Farm — and Hornsmill, on a road of that name, a chamois leather factory between the wars, represents the Domesday mill of Brickenden or of Hertingfordbury. It still contains a fulling plant.

The Lea enters Hertford town by the castle and drove the first mills at the bridge, obvious successors of the Castle Mills bestowed by the Conqueror upon Peter Valognes. They were mentioned several times during the middle ages: in 1247 when William de Valence had custody of them; in 1331 when they possessed two water-wheels; and in 1416 when they were called 'Chastell Mills'. Grinding corn in the early seventeenth century and carrying the name 'Town Mills',[7] (later, temporarily, Islott's Mill), the core of the building could have been of the same date; the mill proper lay back from the line of the street, the buildings on the front, the offices, being later. From the pavement their appearance was of minimal interest but the view from the castle mound could be much more rewarding. Such a vantage point permitted the rear of the mill buildings to be observed and also the bordering gardens, which in blossom time provided a sight little-suspected from the street.

The plant did not seem outstanding in the twenties, the water-wheels[8] having been replaced by a turbine, but the group provided a needed contrasting feature in the 'townscape'. Such sentiment could hardly guarantee its survival and the traffic planners' whirlwind — that has recently struck down so much of old Hertford — has done here likewise, leaving the site a desert, one of those 'void' places the effect of which in towns was condemned by Sir Christopher Wren; and in this instance, having insult added by its use as a car park.

Below this spot the Lea divides, one branch going towards the New River to feed the canal and the other towards the old waterworks site. The miller, by holding back the Lea water, could seriously damage the river connection between Ware and London; for instance, in 1674 William Smith the miller tampered with the sluices, thereby causing a barge to be grounded and nearly destroyed. The Lea, after starting the Lea Valley Canal (or New Cut as it was called last century) wanders past wharves and warehouses until it arrives at a weir at the end of Hartham Lane, which seems to have been a head for the Hertford waterworks. A Malt Mill occupied the same place in the early seventeenth century, while another, which was evidently built here in 1700, achieved local repute for production of a fine quality flour, known as 'Hertfordshire White'. On or near the spot occupied by the last — the 'Flower' mill — the Hertfordshire Waterworks were erected in 1708, though many centuries earlier Philip of Hertford had persuaded the monks of Waltham to improve, for supplying local needs, the springs of Chadwell that were later used by Sir Hugh Myddelton for the New River. The mill, erected in 1708, contained wheels harnessed to pumps but no relic, apart from the weir, survives today.

The ruined walls of Dicker Mill, at the end of the lane of the same name, lie beside the canal but, if it were not for the sound of tumbling water, the site could easily be missed. It was an eighteenth-century structure with weather-boarded sides. It also bore the name 'Lychermill' or 'Priory Mill' and in 1624 was attached to the manor of Hertford. Previously it was evidently the mill used by the priory of St Mary, attached to St John's Church. When the latter was demolished after the Reformation the churchyard was converted into the mill-close, though today the ruins of Dicker Mill are some distance from the situation of the old priory. This can be explained by the removal of Dicker Mill lower downstream in order to give other mills in the town a chance to obtain better supplies of water. When Hertford attempted to restore its lost prosperity in the seventeenth and eighteenth centuries by improving the water connection with London, Dicker Mill was a constant source of trouble; for instance, in 1646 when the miller William Green obstructed navigation by using too much water for his mill.

Three sets of mills provided for the inhabitants of Ware, two at its boundaries and one in the town on the River Lea. The latter site now carries Allen and Hanbury's factory, but in early days its prosperity reflected that of the men of Ware more or less at the expense of 'them' of Hertford; the mill, rented for £5

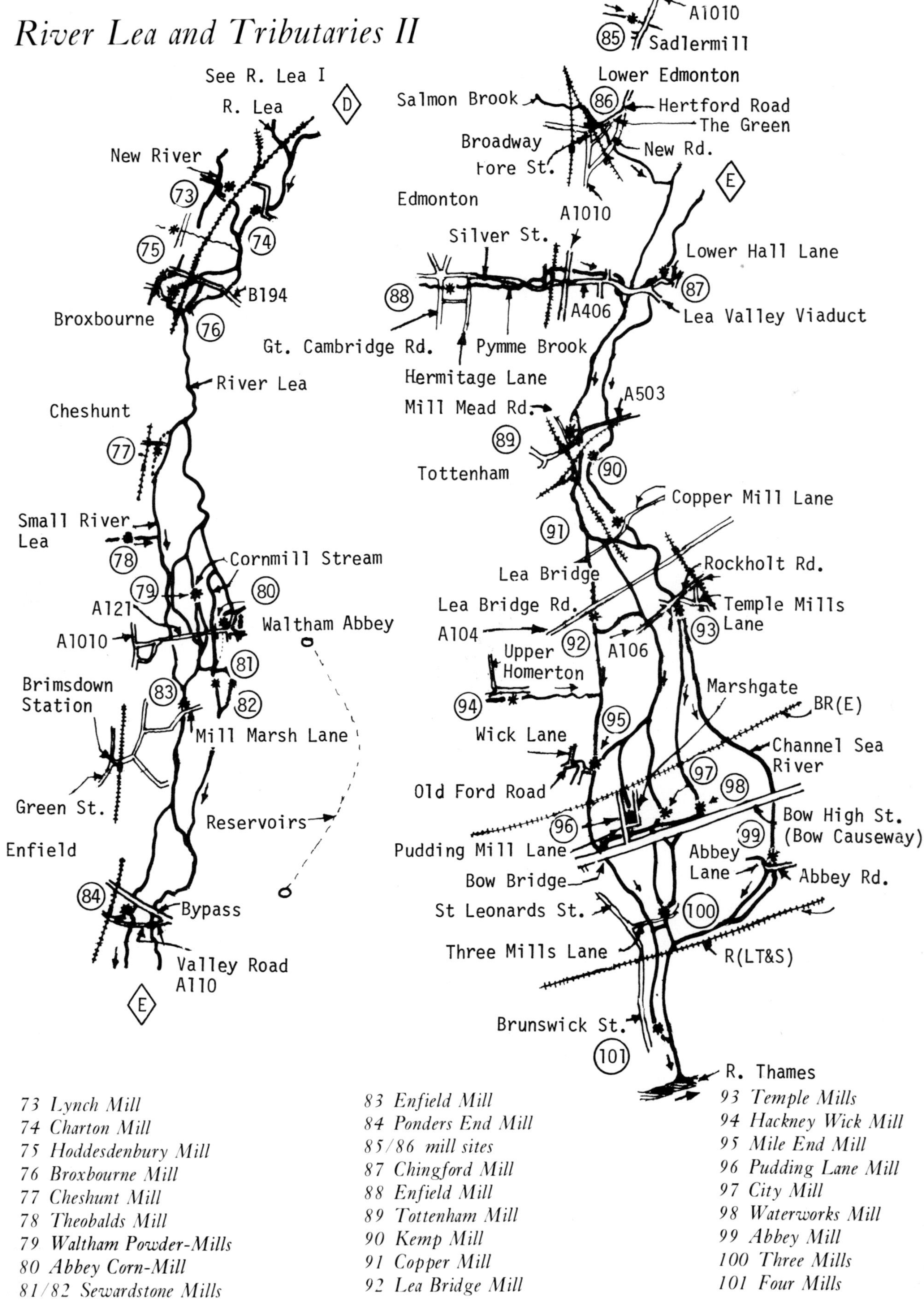

73 Lynch Mill
74 Charton Mill
75 Hoddesdenbury Mill
76 Broxbourne Mill
77 Cheshunt Mill
78 Theobalds Mill
79 Waltham Powder-Mills
80 Abbey Corn-Mill
81/82 Sewardstone Mills
83 Enfield Mill
84 Ponders End Mill
85/86 mill sites
87 Chingford Mill
88 Enfield Mill
89 Tottenham Mill
90 Kemp Mill
91 Copper Mill
92 Lea Bridge Mill
93 Temple Mills
94 Hackney Wick Mill
95 Mile End Mill
96 Pudding Lane Mill
97 City Mill
98 Waterworks Mill
99 Abbey Mill
100 Three Mills
101 Four Mills

in 1420 rose to £28. 13*s*. 4*d*. in 1513. Next would have come Great Amwell Mill, dating from Domesday, which probably stood by one of the weirs half a mile or so east of the church. One of these weirs was reconstructed in 1281 and had damaged the mill, because in 1289 the latter remained out of repair; hence no doubt its disappearance. The Lea flows to Stansted Abbots Mill, once a modern roller-mill apparently given over to another use, and then below Rye House it receives the Stort.

The next mill depends on a tributary but the tail flowed into the Lea. Conduit Lane, once Honey Lane, leads from Hoddesden High Street towards the New River and, crossing the latter, becomes a footpath by the Lynch. Here at the head of a pond below the New River bank stood a mill until the seventies of last century. Of the date of its origin there are doubts. A record of 1569 has the words 'La Lince, where is now built a watermill', but it formerly belonged to 'Dominus Rich', probably Richard Rich the Chancellor who died in 1569. It enjoyed a varied career until its demolition. Robert Cecil in 1603 conceived the idea of conveying water from its mill-pond, which is fed by a spring, to his mansion of Theobalds, but the project was dropped; while two schemes of 1637, one for making the New River more voluminous for the needs of London, and the other, a lottery, for bringing water to the same city, both intended making use of Lynch mill-pool. Owing to the inadequacy of the spring, the mill, about the time of the Civil War, could not work more than ten hours at a stretch, after which four hours had to pass before the pool filled again sufficiently. The miller at the Lynch conceived the bright idea of asking to have the waste water from the New River turned into his pond and Sir William Myddleton, son of the originator, granted him his request. Sometimes when permission was withheld, secret borings were made into the New River bank at the dark of night, so that the wheel of Lynch Mill could turn again. So serious did this nefarious business become that the company had to appoint watchers. Though probably a grist-mill, William Dalby converted it into a cotton mill, yet because the quality of the water damaged the cotton, the scheme failed, and the building was unused until Thomas Bridgeman reopened it as a flour-mill once more in 1799. The inadequacy of the spring prevented the mill from weathering the competition of Victorian times and in 1873 the owner decided to remove it and use the water, which incidentally had never been known to freeze, in irrigating the neighbouring watercress beds. On the mill scales here the famous boxer, Tom King, used to weigh himself, when he was training in the district in 1863 in order to meet the American, John Heenan.

The Lea, level with the Lynch, drove Nazeing's old mill, which stood at Dobb's Weir, and once belonged to the FitzWalter family, but it seems to have decayed, probably because of one of the Lea's periodic floods. J. P. Mauser built Charlton Flour Mills upon the site about 1830 and in one flood lost a team of horses, and nearly a wagoner, who was attempting to pilot them across the ford below the mill-tail. In 1868 Charlton Mills were burned down through the ignition of the flour-charged air, and they were never rebuilt. Before Broxbourne Mill the river receives the Spitel Brook and upstream, west of the main road, stood a mill, probably the example that in 1323 belonged to Agnes de Bassingbourne of Hoddesdenbury Manor, then valued at only 13*s*. 4*d*., because the small volume of the stream in summer only allowed it to work during the winter months. It had vanished by 1500.

Broxbourne Mill, with Broxbourne church on the hill behind, formed a pleasant composition until the last war, but most of the mill has now been cleared away, not that it possessed any feature of interest other than some sixteenth-century brickwork at the rear. As the manorial mill this example would have been given with the manor to the Knights Hospitallers *c*. 1200, who retained it until the Dissolution. Afterwards both were bestowed by the Crown upon John Cock (the same man who received Tewin Manor and mill) and the grant included 'le lokk' through which the supply of water was conveyed from the Lea.

Cheshunt 'Mylle', of Domesday origin, belonged in 1467 to Sir John Clay of Baas Manor, and was then rented yearly at 27*s*., or nine quarters of siligin (a white wheat meal). It stood north of the railway station on the Small River Lea, the western arm. This mill must have had a connection with Perrier Manor, and records state that the latter received in 1594, £1. 6*s*. 8*d*. for permission to divert a stream to Cecil's great house at Theobalds, and on this he probably raised a watermill (to obviate using for his establishment the soke-mill) as both Cheshunt and another mill are mentioned in 1682. Cheshunt Mill probably disappeared soon after 1805 when the Board of Ordnance purchased it for its water rights, to use the latter in connection with famous gunpowder-mills on the Old River Lea, a mile downstream.

At this level the Lea approaches Waltham, the river dividing into three; the Cornmill Stream, Old River Lea and another; the last two flowing through the Powder Mill site, now a research establishment, and on land which was originally Waltham Abbey property, low lying, intersected by many water channels, and covered with much scrubby wood suitable for good charcoal (also anti-blast). Waltham Powder Mills were producing powder as early as 1561 and then covered a fair area. Their progress as the years passed can be gleaned from the words of Thomas Fuller, the seventeenth-century parson, and author, who also happened to be curate of Waltham parish. 'More (gunpowder) is made by mills of late erected on the River Lea betwixt Waltham and London than in all England beside . . . It is questionable whether the making of gunpowder be more profitable or dangerous; the mills in my parish having been five times blown up within seven years but (blessed be God) without the loss of any one man's life.' Despite this assurance, fatal accidents did occur, the first in October 1665, as a notice in the burial register gives: 'Thos Gutridg killed with a powder mill ye 4 day . . . Edward Simons carpenter so killed ye 5 day . .'. In 1672 the owner of the mills was Ralph Hudson, who at a Court Baron, was ordered to demolish them as a nuisance to the folk of the surrounding country. He did not comply with the order, finding it quite profitable even after paying the fines demanded by the manorial court. Fatal consequences were the result of an explosion in 1720, when an entry in the burial register of November of that year says that 'Peter Bennet of ye town killed at ye powder mills ye 27 day'.

Forty years later the mills were considered the premier examples in the kingdom and were then owned by John Walton, a relative of the author of *The Compleat Angler* and a 'Gentleman of known honour and integrity' as put by a contemporary report in the sometimes uninviting phrasing of that period. Farmer's *History of Waltham* gives a view of the works as they appeared in 1765. It shows the river running across the picture past a series of mill buildings, while the artist, intent on demonstrating the rurality of the situation, has drawn a skittish cow, jumping about in the fields. A list of buildings is given which cannot be easily identified, but which include four horses and two dumb mills, corning and glazing engines, one cole and two stamping mills, composition dusting and loading houses, a watch house and counting house, drying leads, charging house and sun stones, etc. About eight sheds lie alongside the stream, then comes the first mill with an undershot wheel overflowing into a pool connected to a canal, as a masted ship is shown therein. Then a house, six more sheds, and the next mill, a double-gabled edifice with an undershot wheel nearly as high as the mill itself, depicted as throwing out vast quantities of spray and bubbling water. Evidently these two examples were the only points of waterpower, the other mills being worked by horses. The same year (1765) an explosion occurred without fatality and it must have been after this that they were rebuilt on a large scale. In 1787 these mills were sold to the Crown by John Walton and reorganized under the superintendence of Sir William Congreve, the engineer.

The Cornmill stream introduces Waltham Abbey, Harold's noted foundation, whose visual magnificence dominated the scene as its social-economic power did its estates. A small town grew around its gates, with a tenantry numerous enough to make the Abbey Mills yield a lucrative rent. These examples stood near the west tower, but today the mill-race and sluice gates alone mark its position. The water comes from the Cornmill stream which leaves the Lea a mile further north. It was an artificial cut made to fill the moat in the abbey precincts and was originally continuous with the western side. The builder is thought to have been Tovi — standard bearer to Canute. He also built the first church in the early eleventh century within his moated homestead and endowed the attached priests.

About that time he diverted the stream to drive a mill for supplying the needs of his household. The earth excavated from the new channel was used to fill the portion connecting the moat and the Cornmill stream. By 1108 this mill was in the possession of Queen Maud and valued at 30*s*. a year. This good woman gave it to the abbot of Waltham in exchange for some land in Aldgate, whereon she founded the celebrated priory of Holy Trinity in that part of London. The convent of Waltham made a profitable asset out of the mill, the townsfolk being obliged to use it. When the abbot leased it in 1528 to a miller named James Blount he obtained the goodly sum of £26. 13*s*. 4*d*. as yearly rent. Like many other spoils of the abbey, this mill, and probably that at Sewardstone as well, found their way to Sir Anthony Denny, whose widow Joane still possessed them in 1553. The mill passed through the hands of many lessees, and then the Government, convinced of the greater importance to the country of the gunpowder-works upstream, purchased the Abbey Mills for their water rights.[9]

The Lea next drove Sewardstone Mill (a blue- and a fulling-mill near each other),[10] whose name is recalled by Mill Marsh Lane at Brimsdown Station. Together with the surrounding marshland and fields it disappeared under the waters of the King George Reservoir. The site is as old as Domesday. The mill once belonged to Waltham Abbey and, after the Dissolution, like those upstream, it came to 'Joane', wife of Sir Anthony Denny, and in the early eighteenth century gave way to the establishment of a calico-mill which enjoyed great prosperity, until the twin disamenities of polluted water and far too free trade caused it to close down. An Act of 1767 for improving the Lea navigation was careful to ensure no violation of the water rights of the various mills on the Lea. It ordered Sewardstone Mill to be supplied with water for washing and cleaning the linens from one of the new proposed cuts, through an orifice six inches in diameter, cut in metal; the idea being to ensure that no more than a certain amount of water could be obtained from the Lea.

Hereabouts, possibly on another branch of the Lea and within the Royal Ordnance Factory area, stood a water-wheel, apparently Crown property but of uncertain antecedents, except that it once drove an oil-mill. In 1804 the Crown used it as the original prime mover for the ordnance factory then started on this site. The Small River Lea continued through what is now 'factory land' to Ponders End. Domesday records a mill for Enfield, which was let to Sir John Wrothe in the sixteenth century for the nominal rent of 6*d*. a year, a day and a half's wages for a labourer at that time. Records for 1572 mention a stream 'coming out from the Lea upon which was anciently a mill belonging to the manor of Enfield'. This can be identified with the present milling concern of Messrs G. R. Wright & Sons near the lock at Ponders End. The nucleus can be recognized at once, its weather-boarded front shows it to be from *c.* 1830. After the First World War the immediate vicinity provided a great attraction in contrast to the wider surroundings. In fact, the many trees alongside the tail framed the older portion and hid somewhat the extensions of more recent years. To a degree this still applies, though much of the tail and pleasant bridge had to give way before the need for lorry turning space, and a corrugated iron hood on the elevation rather emphasizes the conflict often arising between need and looks. Now power comes from electric current, not the Lea waters, and all of the older plant has had to be removed, but nevertheless the approach lane, frontage, with mill and dwellings alongside, convey an atmosphere lacking in other industrial groups.

Next on the Lea would seem to be Chingford Mill, near the Hall. It had closed down long ago and in 1932 it was represented by a shabby Victorian structure with arches under, where the tail flowed. Before mentioning the next site (for sites only remain) Edmonton had a mill, not on the ever-available Lea but on either the Salmon or Pymme Brook, which join it hereabouts. Of Domesday date at least, it no longer survives. Rocque[11] shows a building, though he does not name it, over the Salmon Brook or on a branch near the main northern roads, the Broadway and The Green, Lower Edmonton; the bridge nearby being called 'Saddlers Mill Bridge' *c.* 1800. The other alternative could be at the weir on the Pymme near the arterial road (A10), because a map referring to property in this parish settled by an Act of Parliament in 1801-2,[12] shows a watermill and windmill together in this last position belonging to Sarah Huxley's executors, and in a way this is confirmed today by the presence nearby of a Millfield House. Perhaps both examples existed.

The Lea next comes to Ferry Lane near the Hale Station where, before road widening removed it, could be found Tottenham's Mill. Though not mentioned in Domesday it was held in 1305, shared with other manorial tenants, by Robert de Brus, whose son was so famous. Later inquisitors mention it and another in 1374 stated it was in ruins, with no custom and therefore no rent. A map of 1619[13] showing the other 109 separate houses in Tottenham, also shows this example, quite clearly, spanning the stream, single-storeyed with a vivid red roof. During the eighteenth century it became a paper-mill. However, in 1770, Mr Edmund Wyburd obtained from the owner, a city alderman, a fifty years' lease over the head of the tenant, Thomas Cooke, an excise officer, who evidently conducted the paper manufacture. Perhaps morally justified, he refused to give up possession, greatly to the discomfort and annoyance of Wyburd, who conceived of a good ruse to obtain possession. How he managed it history remains silent, but he removed the roof of the mill, thereby preventing its use. Naturally this exasperated Cooke. Shortly afterwards he met Wyburd when walking on the banks of the Lea. Angry words passed between them, and Cooke taking Wyburd by the 'breech' threw him headlong into the river. As would be expected, a lawsuit followed, which ended in the latter getting his right of possession. He turned the concern into a corn-mill but in 1788 it was burned down and rebuilt immediately afterwards. When the corn- and oil-mill was sold in the next year, an account of the auction gave the amount of grain produced: six pairs of stones ground 300 quarters of wheat in the months of winter, and 150 in those of the summer.

At this time Tottenham Mill then appeared three storeys high, weather-boarded, its two parallel roofs cut by a gable in the centre, several dormers in the roof and a large arch spanning the tail. On another branch of the Lea, not far away from the last example and to the south-east, stood Kemp Mill.[14] Though not shown in the map of 1619 mentioned above, it would be the oil-mill of *c.* 1800, but the making of the neighbouring reservoirs has obliterated the site.

Copper Mill Lane, Walthamstow, amid water-works leads to the site of a mill, which according to Domesday was attached to the manor of Walthamstow Toni or High Hall. Though the manorial corn-mill, it had been converted as early as 1742 to oil crushing. It was made into a copper-mill in 1807 when the ingots were transformed into sheets and utensils of various kinds stamped out or spun. Part of the building housing these processes (but since 1860 serving the waterworks) can be seen, a welcome relief to the technological desert that has in the past fifty years engulfed this valley. The Lea then came to a corn-mill on the south side of the Lea Bridge Road. Known in the seventeenth century as Wadd Mill, its trade was lead manufacture and the surrounding meadowland came to be known as Leadmills Common. The mill, rebuilt in 1833, harnessed its water-wheels to water-pumping machinery to serve the surrounding districts, under the aegis of a certain George Montgomerie.

Hereabouts, near Hackney Wick, the Lea was fed by a small tributary rising in Hornsey, called the Hackney Brook. Where it crossed Wick Road stood a watermill. A survey of the Earl of Cleveland's property in 1652, describes it as a 'Gunpowder Mill . . . adjoining a Pawne Brook', and later the same survey also speaks of 'Pawnes, otherwise ponds formerly kept up for a Mill Pond . .'. The mill must have been rebuilt for it was used for silk, and in 1756 stood at the head of a wide pool, though by 1811 it had been converted to steam power.[15]

The Lea divides again, though the advent of the canal complicates the pattern of the waterways which formerly provoked Leland to remark: '. . . for there be divers socours of streamletts briking out of three principalle parts of Laye Ryvver'. The first site is that of the historic Temple Mills amid railways and factories. This double mill was driven by a branch of the Lea and belonged to the Knights Templars, holders of a manor in Hackney. The Templars were followed by the Hospitallers, who retained possession until the Dissolution, when the mills fell to the Crown. The last ecclesiastical tenant was John Mustyan and a survey and report of 1541 mentions the mill and lists the internal machinery.

> And from £11. 6. 8 from John Mustyan, for the farm of the two mills called Rockeholt, otherwise Temple Mills, situate and lying under one roof within the parishes of Hackney and Leighton in counties of Middlesex and Essex . . . And the aforesaid farmer (J. Mustyan) and his assigns shall support all charges, ordinary and extra ordinary whatsoever due of issuing from the aforesaid mills during aforesaid term. And they shall repair, sustain and maintain the said mills, with their appurtenances, at their own proper costs; to wit, in tymber-worke, coggerounds, shrouds, ladells, trundells, and with stones and generally in all other repairs as often as shall be needful and necessary during the said term; excepting the timber, which shall at all times be supported at the charge of the said late Prior and his successors, as in the same indenture appears.

The appellation 'Rockeholt' is derived from that of a manor house nearby in Leyton, but the name had changed again by 1549 to 'Crathes' mill. The Lansdowne MSS. refer to one of the millers here in connection with certain robbers who, about the year 1599, disguised themselves with false beards and frequented Longton Heath, Leyton and Snaresbrook, where they evidently met together — 'And sumtimes, sum of them ride over by Temple Mills, where I pray you take likewise secret order with the miller that he woulde keepe his gate shut up in the nighte . .'. The writer, Sir Robert Wrothe, had written the above to Lord Burghley's secretary, Michael Hickes.

Temple Mills, after the Civil Wars, were fitted up as boring-mills for cannon, and in the foundry here Prince Rupert carried out his researches. After the Restoration he was made Governor of Windsor Castle, but his duties there did not prevent him from spending much time at these mills. He invented a new gunmetal called 'Prince's Metal' but apparently its secret composition died with him. Afterwards the power of three water-wheels was adapted to grind corn, to bore tree trunks for water pipes, and to cut, grind and polish pins and needles. At a later date the mills were adapted for use as a brass foundry, the management of which in 1721 was alternately assailed and defended in a veritable war of pamphlets. In 1814 they manufactured sheet lead and by 1832 had been converted to flock making. The watermill itself (the site being near the present Eastway–Rockholt Road) was constructed chiefly of wood, and spanned the stream adjoining the 'White Hart', a noted resort of anglers. It disappeared in a most casual way for so well recorded a building, to provide a site for a pumping station.

The massive embankment of the Northern Outfall Sewer crossed the river at Old Ford and here doubtless lies the site of the 'Landmilnes' held by Stephen Asswy in 1344 under the manor of Mile End. The same mill can be identified with one mentioned in 1354 as adjacent to a branch of the Lea. It was leased to a Stratford baker, one Adam Smale, who during rainy weather caused damage to the streamside property of several citizens and also flooded Old Ford Common by stopping the course of the river with 'mud, turf and water doors'. The watermills shown here in old maps disappeared, no doubt, when the Lea was made into a canal. From the time of the Conquest, West Ham and Stepney possessed an array of watermills, of which only one now survives. The desolate surroundings of today would hardly have evoked Camden's comment in 1586, when he describes the Lea as 'divided into three streams, it washes the green meadows and makes them very charming . . .', while Fuller mentions the mill and a neighbouring one as adjoining several acres of land planted with willow trees.

Taking one of these streams, it led in 1932, before the irreversible transformation of this part of London had been completed, to a mill not far from Bow Bridge at the corner of Marshgate and Puddingmill Lanes, in a situation that required a vivid imagination to envisage a long ancestry. This Victorian building, of unredeemable ugliness, standing among slum and squalor, had been demolished and the original watercourse filled in. A predecessor belonged, *c.* 1310, to one John Rickman who gave it to the hospital of St Thomas of Acon in Cheapside.

The alterations which accompanied the erection here in 1932-3 of new bridges make difficult the identification, without the aid of old maps, of the two watermills which belonged to the Cistercian abbey of Stratford Langthorne, called City Mill and Fuller's (or Spielman's) Mill. Of the first Fuller describes it as being attached to several acres of land, planted around with willows, but such a picture gives no help today for concrete in most of its forms has come to reign supreme for a long way on either side of Bow Causeway.

City Mill and the adjacent one were mentioned in the foundation charter of Stratford Abbey; one being held by Edwin, son of Algar, and the other by Ulwin the miller; though by 1312 a benefactor had bestowed them, or their rents, upon the Bridge House Estates for the upkeep of London Bridge.

In 1361 City Mill was evidently leased by a certain William of Tudenham who had in his employ a practical miller named William Poggere. The four city assayers of white bread, whose job it was to collect samples of flour in order to make specimen loaves before fixing the correct weight and price of bread for the London citizens, were supposed to make search for this purpose during October from the city markets at Gracechurch Street, Billingsgate and Queenhithe. However, eight assayers, for some obscure reason, chose to go to Stratford in the month of February and entered Poggere's mill. He either stopped the mill or locked them inside it. Consequently he was hauled before the mayor, aldermen and a concourse of the common people. He admitted his obstruction and contempt of the city assayers and was condemned to the hurdle, but happily the sentence was remitted on account of his age and the dubiousness of the charge. Obviously more than meets the eye lies in this, though of such undertones, records rarely tell.

The Bridge House Estates still remained the owner in 1582, for in that year one Biggs, lessee of Temple Mills upstream, had closed up the lade, called 'Bollyvante', leading to City Mill, because the water level in it had tailed his mill. The lord mayor, in July 1582, wrote to Lord Wentworth about the matter which had been taken to court, for the city authorities and bridgemasters wished to retain their water rights in City Mill despite Mr Biggs.[16]

The Lea next arrives at Three Mills, which travellers on the trains to Barking will have noticed as a large and crudely picturesque group of buildings above the water. Among these can be seen the shell of an immense watermill erected in Georgian times. A date panel gives its age — 1776. Three storeys of stock brick carry a roof of very wide span. The adjacent clock tower and factory buildings have been added since the days when Three Mills had field and marsh for their neighbours and are of the nineteenth century. Bomb damage was widespread here and to learn of the former appearance of this complex and of its detailed later history reference should be made to a booklet devoted to the subject.[17]

In the fourteenth century one of these mills, called 'Warewal', evidently belonged to the priory of Christchurch, Aldgate, and about 1344 they were seriously inconvenienced, the road at Stratford flooded and the growing hay in the surrounding meadows damaged. The cause of this trouble came from a sluice, called 'Fouremullelok' that joined up to the mills further down. The level of the water in it was four feet higher than in any other part of the river, thus a back flow effectively 'tailed' the wheel, or wheels.

The last example on this branch of the Lea, Domesday registered like the last, was Four Mills, near to the modern Sun Mills off St Leonard Street, Bow, belonging to the manor of Stepney. In the eighteenth century the mill building stretched across the river. A glance at the present uninviting edifice strengthens the thought that within those walls no piece from earlier predecessors could be expected, but after excavations, not all that long ago, millstones and timbers were found.[18]

NOTES AND REFERENCES

1 Chauncy, *Historical Antiquities of Hertfordshire*, London 1700.
2 The structure may have been older than its Victorian looks suggested, as a late eighteenth-century sketch (British Museum Maps XV 87) shows the twin-arched bridge, the three-storeyed front, the miller's house alongside, and one of the great chestnut trees
3 Information kindly supplied by the late Mr E. A. Mardon
4 Fisher, J. L., *The Deanery of Harlow*, Benham and Sons, Colchester, 1922
5 Information from notes by Mr N. T. Bagshaw
6 Chauncy, op. cit.
7 Speed, J., *Theatre of Great Britain*, Sudbury and Humble, London, 1611
8 Shown in Andrew Wren's map of Hertford, 1766
9 French, J., 'Moated Grange and Mill at Waltham Abbey', *Essex Archaeological Society Proceedings*, vol. X, 1908, p. 541
10 Cary, *Map of Middlesex*, 1786
11 Rocque, *Map of Middlesex*
12 Hodson and Frost, *History of Enfield*
13 Robinson, W., *History of Tottenham*, Tottenham, 1818
14 Cary, *Map of London*, 1819; Rocque, *Map of London*, 1745
15 F.R.C.S., *Glimpses of Ancient Hackney*, Hackney Mercury Office, London, 1894
16 Gregg, W. W., *Remembrancia*, Malone Society, London, 1907
17 Gardner, E. M., *The Three Mills, Bromley by Bow*, SPAB Booklet No 4, London, 1957.
18 See the *Evening News*, 4 May 1950

Three Mills clock tower
(1929)

Chapter 17

THE RODING AND CHELMER, AND SOUTH ESSEX

RIVER RODING

THE first mills to be driven by this stream would probably have been two belonging to the manor of Great Canfield, near Dunmow, obtained in 1579 by John Wiseman from Edward de Vere, Earl of Oxford. Perhaps they can be identified, though the sites cannot, as being attached to High Roding in 1562. The first certain site was Waples Mill, Berners Roding, weather-boarded with a roof covered in corrugated iron. The iron water-wheel had been removed before a visit in 1931 when the owner said that he had replaced the existing boarding during the previous thirty years and, although the mill was only used as a store room, he did not feel it right to demolish it. However, today only a sad reminder remains in pieces of timber, a post here, a truss there, lying around on the site. The next mill, south of the village of Fyfield, was once approached by an avenue of poplars alongside the pool. However, Fyfield Mill, though at one time virtually derelict, still stands, and has been restored to working condition driven by a turbine. The water-wheel has gone, the water seeps through the wheel-race walls, but it is still an appealing little building with its weather-boarding and archaic pedimented windows.

The Crispey Brook, which joins the Roding at Chipping Ongar, supplied a watermill at Pedlars End, Moreton. It did not survive last century, nor was it shown on Chapman and André's map.[1]

Chipping Ongar, despite its importance in earlier times, seems to have possessed no mill of its own. Perhaps those at Langford Bridge to the south and in the adjacent parish of Kelvedon Hatch, which disappeared last century, served instead.

The surroundings of Littlebury Mill are in many ways much more pleasant than the mill itself. A large pool can be remembered, which remained until 1947 when the water was still used by the mill for powering its water-wheel, but now this nineteenth-century building is used only as a store.

Navestock Mill which stood at Shonks Mill Bridge, the last of its line being destroyed about 1890, is mentioned several times. First in 1222 when it was worked by one 'Geoffrey the Miller' for the canons of St Pauls, and in 1356 when the profits maintained a light before one of the altars in St Paul's Cathedral. The donor of this bequest was John Barnett who in that year exchanged his prebendary at Lichfield for the archdeaconry of London. He died in 1370, after being bishop of Worcester, treasurer of England and bishop of Bath and Wells.

Stapleford Mill, standing near the recently rebuilt Passingford Bridge, is well known for its own beauty and surviving unspoiled surroundings. This mill is not only excellently proportioned and shaped but is also accompanied by the right amount of farm building, and links the road to the river, the group providing a satisfactory ensemble. Few additions have accrued to the original building so that the latter can be seen much as when first built. The structure dates from an early 18th-century rebuilding with a plain pitched roof.

The water-wheel, visible in the late twenties, gave way in 1931 to a turbine, at which time apparently the owners removed the wooden machinery and installed the present plant, still used for coarse grinding. Messrs Twynham, the owners, have commendably taken care of their mill, though its future, owing to the mounting cost of repairs, causes concern.

Further downstream a rivulet from Epping joins the Roding and near the bridge stood in 1588 'Garnnishe Mill', then a tenement belonging to the manor of Theydon Garnon. In 1800 the surrounding buildings were called Gernon Mill Farm.

The Roding passes under a bridge near Buckhurst Hill Station and near here must have stood Chigwell Mill. The approach to it was evidently by the causeway which also acted as the dam. In 1364 this dam failed in some way with two results: either the water flowed

Waples Mill, Berners Roding, Essex (1931)

Moulsham Mill, Chelmsford (1931)

Stapleford Mill, Passingford Bridge (1929)

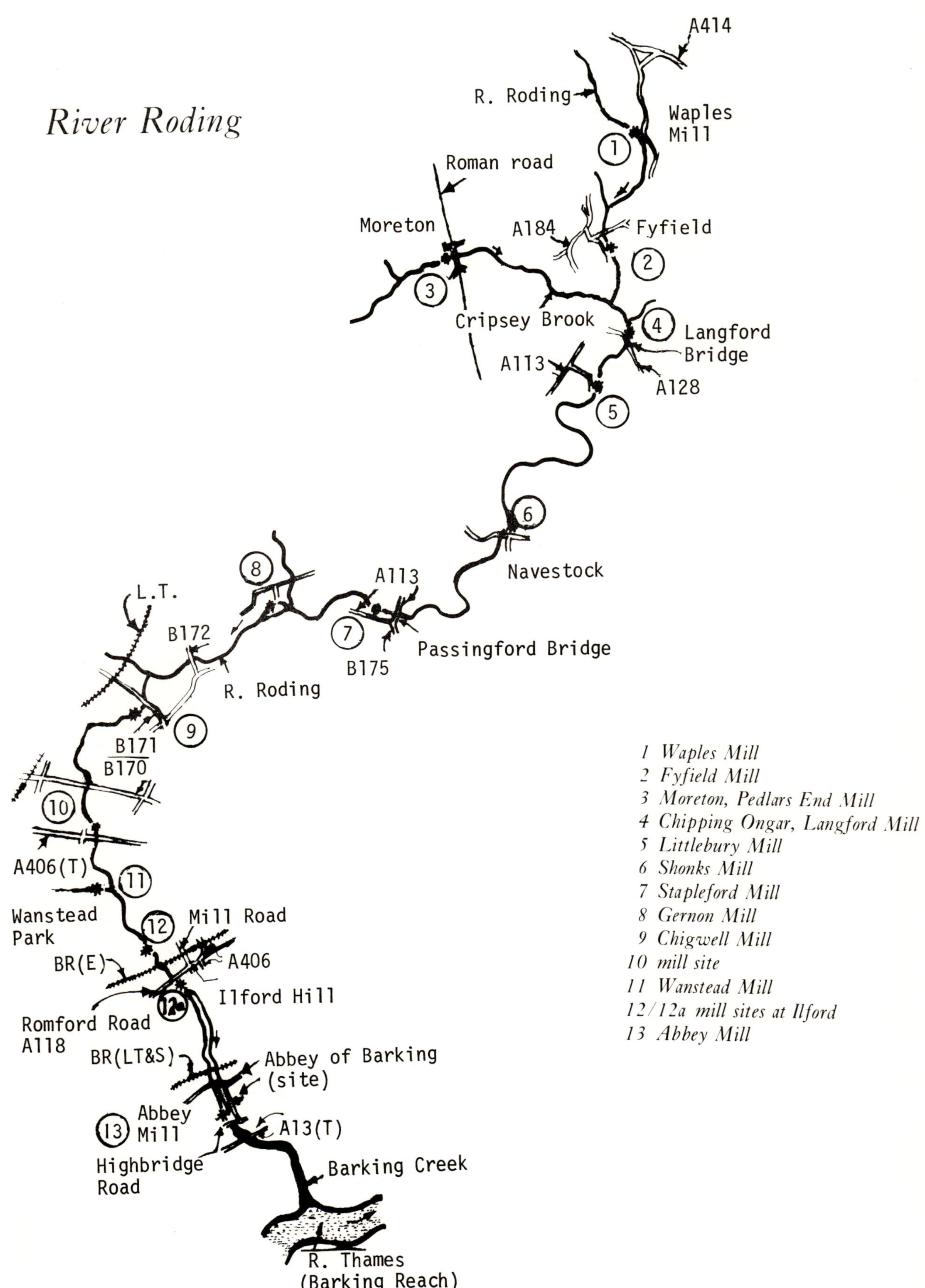
River Roding
A414
R. Roding
Waples
Mill
1
Roman road
Moreton
A184
Fyfield
2
3
Cripsey Brook
4
Langford
Bridge
A113
A128
5
6
Navestock
8
A113
L.T.
B172
7
Passingford Bridge
B175
R. Roding
9
B171
B170
10
A406(T)
11
Wanstead
Park
12
Mill Road
BR(E)
A406
Ilford Hill
12a
Romford Road
A118
BR(LT&S)
Abbey of Barking
(site)
Abbey
Mill
13
A13(T)
Highbridge
Road
Barking Creek
R. Thames
(Barking Reach)
1 Waples Mill
2 Fyfield Mill
3 Moreton, Pedlars End Mill
4 Chipping Ongar, Langford Mill
5 Littlebury Mill
6 Shonks Mill
7 Stapleford Mill
8 Gernon Mill
9 Chigwell Mill
10 mill site
11 Wanstead Mill
12/12a mill sites at Ilford
13 Abbey Mill

over the top and endangered the safety of anyone who wished to take grain to the mill, or insufficient water would remain in the pool, so that the mill could 'neither grind nor be of service to the community'. In this case the Crown charged a London citizen as being bound to repair the pond, etc., because he held land next to it. This good man, Nicholas Plunket, denied responsibility saying he was only bailiff to the lady of fee, Alice de Perrers.

This took place before this lady's unpopular association with Edward III. And this occurrence incidentally shows her as a lady of some estate not, as some of her detractors whispered, a 'tiler's or weaver's daughter' nor 'a domestic drudge'.

A small tributary of the Roding before it joined the main river must have driven the watermill at 'Wanstede' possessed by the nunnery of Clerkenwell. It was probably situated near the 'Melgrave'[1] of medieval times. Rocque does not show it in his map[2] and as there is no trace of its location today no doubt the making of the famous and extensive ornamental waters at Wanstead Park sealed its fate. Likewise no signs are evident at Ilford of the sites of watermills attached to the manors of Wyfields and Uphall.

A watermill could hardly be expected to be found at Barking but the retained foundations of the abbey walls, and the nearby parish church, are witnesses to this place's past. A little to the south-west actually stood the convent mill which was pulled down in 1922 but the site can be found by Granary Wharf at the end of Heath Street. Domesday records it and it was the property of the noted abbey and was leased. In 1243 it appears that the abbess closed down, by friendly agreement, 'the rival mills of William and Godfrey Dunn' at Barking. It does not seem clear whether these were buildings alongside the main Abbey Mills or were horse mills or possibly some other windmills elsewhere but by 1260 records say that oaks were given to repair the convent mills.

The mills still worked in the early nineteenth century and were then the property of Sir Edward Hasle who bound the miller by an agreement to grind the grain for the poor at a nominal charge. The mill, demolished some time ago, left only the miller's Georgian house, but this has now also gone too. Fortunately a sketch of about 1800 gives an idea of what Barking Mill looked like at that time. This illustration[3] shows the creek full of ships; houses, the quay and several warehouses, long since replaced. It also shows some fine elms on the land between the two bridges, and the mill itself of wood with three gabled roofs. With St Margaret's Church in the background, it provided a most attractive picture.

RIVERS IN SOUTH ESSEX

Next to be considered are several sites, of which some can be identified, on the streams that drain into the Thames, and on the tidal creeks formed from the same river where tides provided the power. On the Weald Brook, by Purtwell Bridge near Brentwood a watermill existed as late as the last century. Five hundred years earlier it belonged to the Spital which stood opposite on the north side of the road. This mill had become alienated from the hospital in 1514 and was attached to the manor of Ropers, and named from that noted family.

A considerable distance further on towards Rainham village the Ingrebourne Brook, as the stream had come to be known, can be seen corsetted between modern banks, or banks heightened in recent times, but alongside the Bell Inn in this village the present much befouled stream could represent the lade to serve the mill once hereabouts. It would have vanished long ago, and no relics are there to indicate firmly, as old maps do not show it, what may be only a probability. If so, it would have been the sole watermill for the widespread liberty of Havering, comprising Romford and Hornchurch, which was attached to the manor of Dovers, situated near Rainham village. Alice, wife of Richard de Dover, owned it in Henry I's time and it remained in that family until 1335.

The Knights Hospitallers held a tide-mill at Purfleet upon the Mardyke. Until the eighteenth century the tide flowed up the Mardyke as far as Bulphan, inundating the fields and leaving a trail of rubbish and other undesirable detritus. This much bewailed nuisance, and its proximity to the important Tilbury Fort, prompted the Board of Ordnance to purchase it. A quay, houses and barracks were erected nearby; the mill was demolished, and on its site a powder magazine was erected to provide an alternative to the storage which then lay at Greenwich. As the watermill no longer remained, no need arose for the tide to fill the reservoir. A wall was built across and the complaints ceased. It is almost impossible today to locate even a probable site. The appropriate accompanying earthworks are missing while, near Purfleet, the bridge built in 1874 and the concrete walling of that time to contain the stream, make any recognition less likely.

In the south-east corner of Essex records name several mills. As most of them date to at least Domesday, windmills can be ruled out and although horse-mills and cattle-mills could have been a possibility the motive power would have been more likely the tides.

An exception was that watermill driven from a large pond in Belhuss Park, Aveley. In 1341 it was held by the clerk, Nicholas Staunton, one of those civil servants so important to the Crown's administration at that period. It vanished long ago but the banks of the mill pond and the mill dam remain quite evident today. South Ockenden retains its mighty moat, at the south-east corner of which runs the overflow. Hereon stood a watermill. Today no sign other than the flowing water reminds one of its former presence. The two mills owned by Barking Abbey, one at Corringham and the other at Mucking have unknown sites. These would, in all probability, be tide-mills but the massive topographical unheavals in this area since last century have made searching for the sites unrewarding. However, these can be suggested. For Mucking, just north-east of the church near where an embankment crosses a depression in the marshlands, and, for Corringham, at Island Mill off Shellhaven Creek, or by Oil Mill Creek off the Saltfleet.

Vange, Pitsea and the Benfleets had mills in Domesday, with sites not identified but probably on one of the many creeks of the Thames. Hadleigh had its mill from Domesday until the fifteenth century. The 'Milflete', shown on old maps as late as 1801, formed a wide navigable creek in those early times and near it, so it is thought, a pond which stored tide water supplied this mill. On the other hand just east, below the ruined castle, the ground forms a sharp defile crossed for a part by an earthen bank and although this piece may be somewhat shapeless today, it could have supported the watermill. The catchment area would seem to have been sufficient only for a small mill of intermittent use, a possibility for one that might have served only the castle garrison.

At Battlesbridge can be remembered some picturesque mill buildings long associated with the site. Here again part-tide and part-river (Crouch) provided the motive power. A fire in 1932 destroyed all but a small portion of those buildings. Before then, steam power had usurped that of the tides. Today it is hardly worth a visit. On a neighbouring creek, where the River Roach becomes tidal, could be found alongside a giant roller-mill, Great Stambridge Mill of eighteenth- to nineteenth-century construction; yet, with the removal of the water-wheel in 1947 and a fire a few years ago, little more than a wreck remains. Little Warley, near Brentwood, in early records was sometimes given the name 'De Septem Molis' and 'Septem Molar' because it once had seven mills. Why it may be asked, because today not a vestige remains and Domesday keeps silent on the matter.

THE CHELMER VALLEY

The first, that at Tilty, once could have been considered the best that remains. At the right time of year and day, a place more pleasing than that below the ruins here could not be found. The Cistercian pioneers always chose beautiful situations, and a typical one is this where they built St Mary's Abbey of Tilty. In 1133 Maurice Fitz Geoffrey chose this land to found his monastery, bestowing all his property in the parish upon it including the watermill. It could be seen in the early thirties as an ordinary seventeenth-century building of white colour-washed brick, where the uneven tiled roof suggested older work underneath. It also produced, as watermills do sometimes, an eerie, mumbling sound at short intervals, followed by utter silence. A dribble of water from the pool would fill the top buckets of the wheel, which, when sufficiently weighted revolved very slowly, emptying the water into the tail beneath.

Today a sad deterioration must be admitted; the situation no longer seems remote, a new suburban village has arisen to the east, the agricultural environment no longer shows the same attractiveness and the mill itself is derelict, having been disused for a long time. The cast iron water-wheel remains *in situ* and the mill is perhaps capable of restoration.

Time has removed the evidence of two mills at Great Easton, recorded as belonging in 1597 to a member of the Cromwell Family. Next downstream is Elmbridge Mill. It ceased grinding between the wars and had been much altered. Fortunately in 1940 the then owner converted it into a dwelling, and though most of the external appearance, weather-boarding and tiles, is not of great age, its shape remains; at the time the plant was removed.

The next parish, Little Dunmow, of 'Flitch' fame, once contained the abbey, founded by Juga, sister of a local magnate, Ralph Baynard. Situated on a prominence it would have been conspicuous amongst the surroundings. Its visual impact on this stretch of Essex landscape has been replaced by that of a giant sugar-beet factory arising in the vicinity. A sunken depression near the ruins, that has a stream flowing through it, was either the convent's mill-pool or fish-pond. A 'Peper Mill' recorded for some spot near Great Garnett's mansion, would have been on a tributary entering south of Dunmow but like so many others the site is uncertain. A watermill called Poole Mill stood south of Great Dunmow in 1602-3.[4]

Fortune has favoured Felsted Mill, as when it ceased working, little hope existed of keeping this distinguished-looking Victorian building (as can now be admitted, though forty or more years ago a different taste would have been hostile) sited so delightfully where the road meets up with the river, accompanied by several waterways and the dominance of many large chestnuts. Removed, regrettably, are the plant and water-wheel but a conversion of the lower two floors to a dwelling has been successfully carried out, and externally provides a model as to how unobtrusive such a work should be.

To judge by details, the same builder must have had a hand in designing Stebbing (Bran End) Mill a few miles further north. Perhaps the present mill (named Absol two hundred years ago) occupies the site of one belonging in 1373 to Leigh's Priory, and also of 'le corne mille' of 1472 appurtenant to Walthambury Manor in Great Waltham, for whose repair four oaks were felled in Absol or Apchild Park nearby.

Hartford End Mill, once one of the best in these parts, belongs to both Great Waltham and Felstead, the boundary lying in the centre of the lade. The group is delightful with its roof of rich red tiles and long façade of plum-coloured bricks pierced by small windows. Since the twenties little elevational

Tilty Mill, Essex (1931)

Elmbridge Mill, Dunmow (1931)

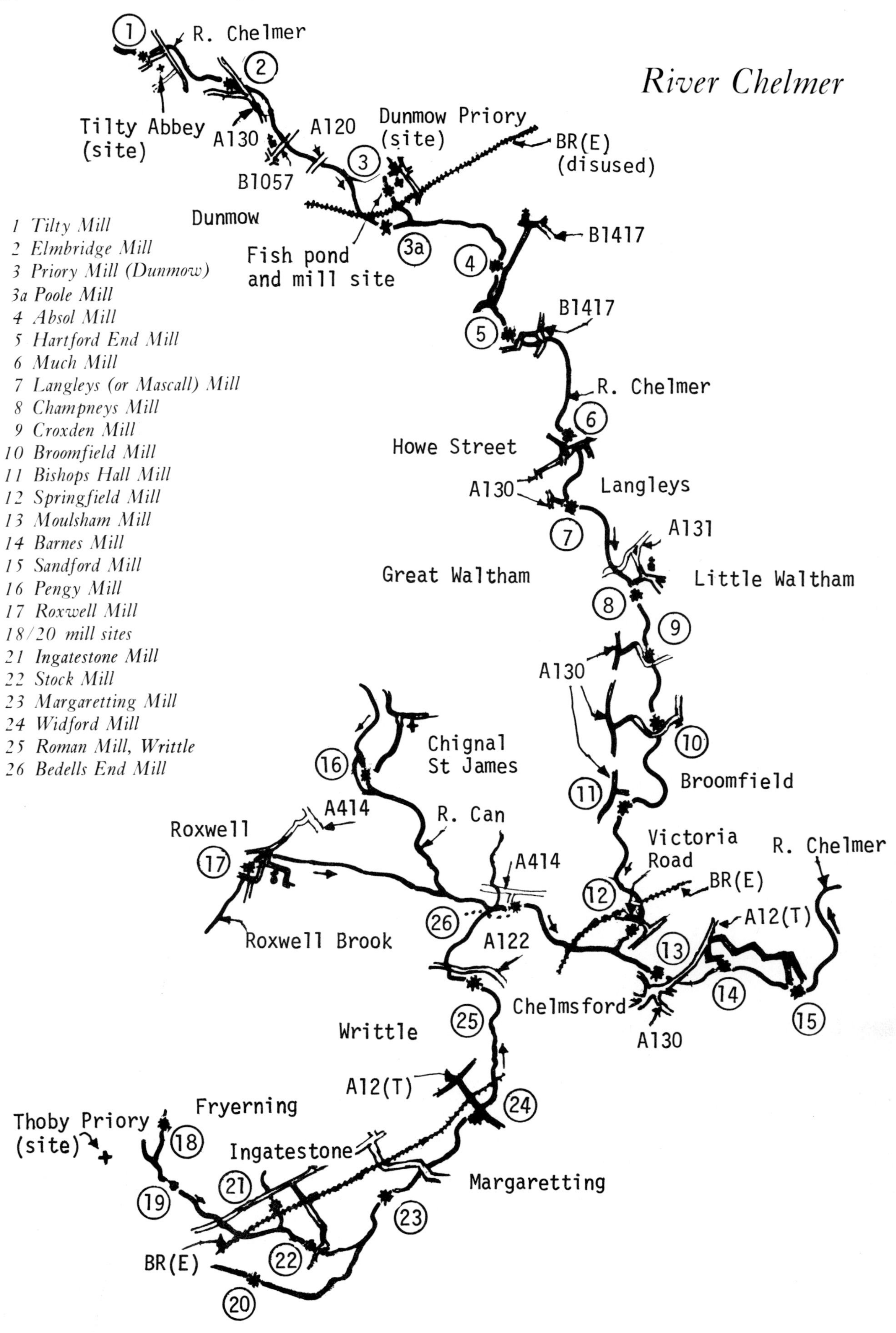
River Chelmer
R. Chelmer
Tilty Abbey (site)
A130
A120
Dunmow Priory (site)
BR(E) (disused)
B1057
Dunmow
Fish pond and mill site
B1417
B1417
R. Chelmer
Howe Street
A130
Langleys
A131
Great Waltham
Little Waltham
A130
Broomfield
Chignal St James
R. Can
A414
Roxwell
A414
Victoria Road
R. Chelmer
BR(E)
A12(T)
Roxwell Brook
A122
Chelmsford
A130
Writtle
A12(T)
Fryerning
Thoby Priory (site)
Ingatestone
Margaretting
BR(E)
1 Tilty Mill
2 Elmbridge Mill
3 Priory Mill (Dunmow)
3a Poole Mill
4 Absol Mill
5 Hartford End Mill
6 Much Mill
7 Langleys (or Mascall) Mill
8 Champneys Mill
9 Croxden Mill
10 Broomfield Mill
11 Bishops Hall Mill
12 Springfield Mill
13 Moulsham Mill
14 Barnes Mill
15 Sandford Mill
16 Pengy Mill
17 Roxwell Mill
18/20 mill sites
21 Ingatestone Mill
22 Stock Mill
23 Margaretting Mill
24 Widford Mill
25 Roman Mill, Writtle
26 Bedells End Mill

alteration has taken place; its owner cares for it well, and with the adjoining house, the group could hardly be bettered.

All the same the interior lies almost bare and lacks the interest of former times. The timber frame, of huge scantling in parts, with both adzed and planed surfaces, and floor-boarding in all sorts and sizes of planks, makes it difficult to determine the age of each and every addition and alteration. The caretaker's house dates from the sixteenth century, as also does the centre portion of the mill with a seventeenth- to eighteenth-century brick facing, while the far portion could be *c.* 1780. A prosperous Victorian miller added on the north-east end a house for himself.

Of the machinery, only the upright shaft and crown-wheel remain, while of the water-wheel there is only the mill-race. This was, when revolving, most impressive, almost startling and claimed to be the largest in Essex, twenty-three feet in diameter and ten feet wide, the axle alone being two feet thick. Prille Mill, a fulling-mill mentioned during the fourteenth and fifteenth centuries in the Court Rolls of Walthambury Manor, may have stood upon this site.

Lower down, near Howe Street, stood the eighteenth-century mill that occupied the site of 'Great', 'Mochel', or 'Muche' Mill references to which occur often in the Court Rolls of Walthambury Manor. The first mention is in 1397, and for over thirty years following it was under constant repair.[5]

Mochel Mill, in 1400, 'being olde and ruinous in respect of its opus acquaticum' was put in repair and remained so until 1404, when further refurbishing was undertaken. In 1408, the manorial officer concerned, the beadle, was directed to repair the dam that was out of order, and again in 1410 similar instructions were given, this time to attend to the roof. About the same date, the manorial court witnessed a curious 'presentment'. Two of the men working at Mochel Mill, John Wolstan and William Meller, were charged with appropriating, without licence, three old mill-stones valued at the then modest sum of 40*d.* The two men claimed it was a custom for anyone to take old stones after replacing the new. Evidently the court was impressed by this defence for it recorded no decision.

The watercourses appear to have been the most vulnerable parts of Mochel Mill, for again in 1412 the beadle had to supervise repairs underneath the mill; that the repairs were under and not at the side could have been illustrated in the last of the line to survive, where the watercourse could be seen passing under the centre. May 1413 saw more general mending, and a new overflow, a precaution probably suggested by heavy rain recorded for the previous April. The beadle was found negligent and near Christmas was reprimanded for omitting to carry out necessary restorations, which he afterwards did but apparently in a perfunctory manner, because in 1415 the flood-gates again required more attention, and the water-wheel needed further repairs. Next year, after being in thorough working order, the mill experienced its old trouble again: the flood-gates and 'opus terraneum' (i.e. banks of the stream) failed once more. No more record of concern until 1426 when 'Mochel Mill' was in 'debili punctu', but it had to wait two years before wood was taken from Littley Park, two miles further upstream, for a new water-wheel.

At this date the miller was one John Wolstan, surely he who had been accused of taking the mill-stones, and who seems to have been prone to remove unconsidered trifles. But this time not mere trifles — boards, collars of iron, hewing blocks, planks and other objects, not unconsidered, for the court fined him 12*d.* An air of irresponsibility seems to have been within Mochel Mill, as witness the servant of John Turnour of Writtle (who also leased Moulsham Mill at Chelmsford). This man, William Palmer, when left in charge of the mill absented himself without reason, and left the key remaining in the outside of the door. Someone made off with two bushels of wheat flour and various other articles, to a loss of 7*s.* 8*d.*, for which Turnour sued Palmer in the Great Waltham Court Baron.

In 1436 the watermill was rebuilt, oaks from both Littley and Apchild Parks being felled for construction of the water-wheel. Many years intervened before the next mention in 1563, when the mill was let to Nicholas Rychards and was described as a 'water myll with cotage coveryd with straywe called Myche Mylle, with the Mylle Meade (1 acre 3 rood 23 pols) and 39 poles of pasture between the mill and the back river'.

Another mill stood on the Chelmer, just close to Langley's Manor House, and was named variously Berwyks, Langleys or Mascalles mill. In 1530 it partly belonged to Richard Everard who held a moiety of Langleys manor. There remains today only the position of the watercourses. The same can be said of the site of Champneys Mill of Little Waltham further south, which vanished many years ago. In the latter part of the fourteenth century, it was fulling cloth, but by 1418 it had been converted to corn-grinding, with a miller named John Abbott. No excerpts from the Court Rolls can tell of the disappearance of this mill, but that its pool contained succulent fish is certain, for in 1418, one 'Stephen atte Lyll', in defiance of the good miller, had persisted on occasions during the two previous years in taking fish from the pool by 'spertes, lammes and schoffnetts'. No indication was given as to the nature of these fearsome sounding instruments.

Croxden Mill, Little Waltham, which lies in the valley, looks picturesque from the main road; the yellow cornfields in the foreground, the grey and russet structure and the dark foliage of the trees on the river banks, form a pleasant harmony. The pool, no wider than the river which feeds it, drove an undershot wheel, which was working in the '30s. In 1404 a new timber bridge was built over the river, but it has long gone, and a ford, disused for many years and overgrown with vegetation, crosses the tail. This example lay derelict for a long time, with the plant and

Writtle Mill, Essex (1931)

Littlebury Mill (1929)

Absol Mill, Felsted (1931)

Croxden Mill (1931)

water-wheel removed. Fortunately it found a sympathetic owner, whose renovation of it retained the internal structure and reclad it with weather-boarding, thus keeping the original shape and more or less its pristine appearance..[5]

External repairs were made to Croxden Mill before the winter of 1407 and in the year following trees were felled for further restorations. The water-wheel required renewing in 1411 but further work was not needed for some years. The miller at this time made a habit of holding back the water, greatly to the damage of the upper pool. The Manor Court in 1415 told him to desist or he would be fined. Jeffrey Dunbed, a miller, took the mill from 1419 for seven years and after his lease had expired a general restoration took place, while the mill dam and machinery connected with the watercourses needed repair in 1430 and 1432.

Broomfield Mill, a mile downstream, was a large example with no particular beauty. Today its site has given place to a garden in which grows a fair-sized thorn tree showing how long ago this particular building was removed. The pen-stocks remain and, as so often happens, the lower wall still shows the wheel-shaft opening now bricked up. Nearer Chelmsford the small watermill attached to Bishop Hall of the same parish has been ousted by the present factory though some years ago the rather meek, Georgian, miller's house could be seen.

One would avoid the visual devastation that makes up present day Chelmsford, and move on to the bypass just east of the town for a look at Springfield Mill. This, if less conspicuous than it seemed thirty and more years ago, still cannot be easily missed. At that time it stood in a quiet lane near the old bridge facing some old cottages. Since then, road widening, the uninspired removal of the bridge, the setting up of unsightly flood control apparatus over the river channels, and the springing up of a multi-storeyed office block nearby have now made Springfield Mill look out of place. Before the First World War a turbine had replaced the water-wheel, which subsequently gave way to electricity. Its present use is as a store.

The date of the building is early nineteenth century and illustrates, as others do, the tenaciousness of the tradition of good design and fit use of materials even in an unpretentious and utilitarian structure like this. The essence of the pleasant appearance of Springfield

Springfield Mill, Chelmsford (1931)

Mill apart from plain materials, such as weather-boarding, brickwork, and tiles, lies in proportion and scale; for instance in the size of voids and their relation to the façade. More good sense on the part of its builder is exemplified in the adjacent Georgian residence, where a bay window projects on to the path-way. This encroachment on the right of way interferes little with convenience, and yet it affords the requisite contrast to the flat façade of the rest of the mill.

The Chelmer Canal starts not far away. It was projected in 1765 but, owing to the dilatoriness of certain persons, a new Act was required in 1793, and even then the opening did not take place until 1797. However, the engineers seem to have done little more than canalize the existing Chelmer. This river is joined by the Can at Moulsham Mill, which looks a little stark across the meadow land as seen from the motorways that have carved up the scenery.

Forty years ago accepted taste insisted that most Victorian watermill buildings be dismissed as at the bottom of the aesthetic scale, but performances of the ensuing years have taught us to be thankful for such lesser mercies. Moulsham Mill, or that half of it visible from the road approach, may seem at least to echo something of the elegance that would be expected from its eighteenth-century predecessor. The latter succumbed to a fire about 1850 and the replacement fits into the scene, due not a little to the use of weather-boarding.

Though the water-wheel provided the power until 1953, it was subsequently replaced by electricity, until the business closed about 1970. Inside can be found the numerous pieces of grain and meal processing apparatus that have been acquired since its building.[6] The great frame made use of pitch pine yet many oak members remaining from the previous building show how late this re-use was practised. Between the spacious Georgian mill-house and mill is a gabled three-storeyed building, whose plaster work may cover a sixteenth-century house or a fragment from an earlier mill. In the sixteenth century Moulsham Mill was purchased by Thomas Myldmay and in 1591 a survey called it 'a very good watermill'. With Springfield Mill it was owned by the Strutt family from 1660 to 1780. Later it came to the Marriages, the noted Essex milling family.

RIVER CAN AND RIVER WID

These streams flow into the Chelmer. On the Can, Pengy Mill could be found near Chignall St James but only a mill-house remains. Another tributary, the Roxwell Brook, fed Roxwell Mill, at the village of the same name. In a sad condition in 1931, looking unlikely to survive many more years, it is no longer to be seen. Also to be seen was the cast iron water-wheel, though not working as most of the machinery had been taken away. The regular framework of the mill showed it to be eighteenth century, with the familiar evidence of older work incorporated, work that may have derived from repairs recorded as carried out in 1399-1400 by one John Clonger. Apparently at a subsequent date a new house on its site incorporated the substance of the old watermill structure.

Bedells End Mill gave way long ago to a conglomeration of industrial buildings. The head stream of the Wid rises north-west of Ingatestone, and by repute drove certain mills at sites not really identifiable: one at Thoby Priory, one at Ingatestone Grange, and another at Furze Hall. Then the manor of Arnolds near Mountnessing held two watermills in 1569 and Stock Mill (near Ingatestone) was held in 1476 by Sir Thomas Tyrell, son of a Speaker of the House of Commons. All these have completely disappeared. The last may have been on the River Wid at Mill Hole in Margaretting, just east of Stepel's Farm.

A more tangible relic, at least until its demolition in 1931, was the watermill just north of the railway, south of Ingagestone itself. Lower downstream, Widford Mill stood between Galleywood Common and the railway line and is to be seen today in the guise of a brick and slate hut. The next site, that at Writtle, no longer repays a visit in the way it did years ago. In summer one could fine in the array of flowers and rock plants, in the lichens and smaller vegetation which had sought out every crevice both in brickwork, and in concrete walling that bordered the spillway and tail, a riot of colour unrivalled at any other watermill in the area, and a feature to redeem the not very prepossessing appearance of Writtle Mill itself. Though it ceased as a mill long ago, it stayed as a store, and in a sense continues so, but the pool, and the related waterways as well as that spontaneous garden, have vanished.

In the eighteenth century some inhabitants at Writtle remembered a 'Roman Mill'. Probably it stood on the site of this one and was attached to the Rectory Manor, which is still perpetuated in the name 'Roman's Fee' given to part of the parish.

The origin of the name is unusual. In early times there had been established in Rome a hospital for the help and maintenance of the infirm sick and poor. It belonged to the English and was called 'The Holy Ghost Hospital [or Hostel] by the Church of St Mary Saxia Rome'; and in 1203 King John, who had his palace at Writtle, gave the church, rectory, manor, and mill, to this institution for its support — hence the name. But this connection must have been broken during the Black Death for in the late fourteenth century William of Wykeham bought the rectory as endowment for his new college of St Mary Winton, Oxford.

RIVER CHELMER

Beyond Moulsham lies Barnes Mill on the canalized Chelmer, its former rural environment almost engulfed by Chelmsford's suburbia. In the thirties this

South Essex

1 Spital Mill
2 Rainham (Dovers) Mill
3 Aveley Mill
4 Ockenden Mill
5 Mardyke (Tide) Mill
6 Mucking Mill
7 Oil Mill
8 Pitsea Mill
9 Castle Mill
(6 to 9 probably tide-mills)
Weald Brook
Brook Street
BR(E)
A12
South Ockenden Hall
Moat
BR(E)
A125
A13(T)
Ingerbourne R.
BR(E)
Rainham
B1335
Belhus Pk.
B186
B1335
Aveley
River Thames
A13
The Mardyke
A1090
Purfleet
Pitsea
Pitsea Hall Fleet
BR(LT&S)
Hadleigh Castle
Mill flete
Vange Fleet
Bowers Marshes
Canvey Island
B1420
Corringham
A1014
Holehaven Creek
BR(LT&S)
Oil Mill Creek
A1014
Mucking
Shellhaven
River Thames Estuary

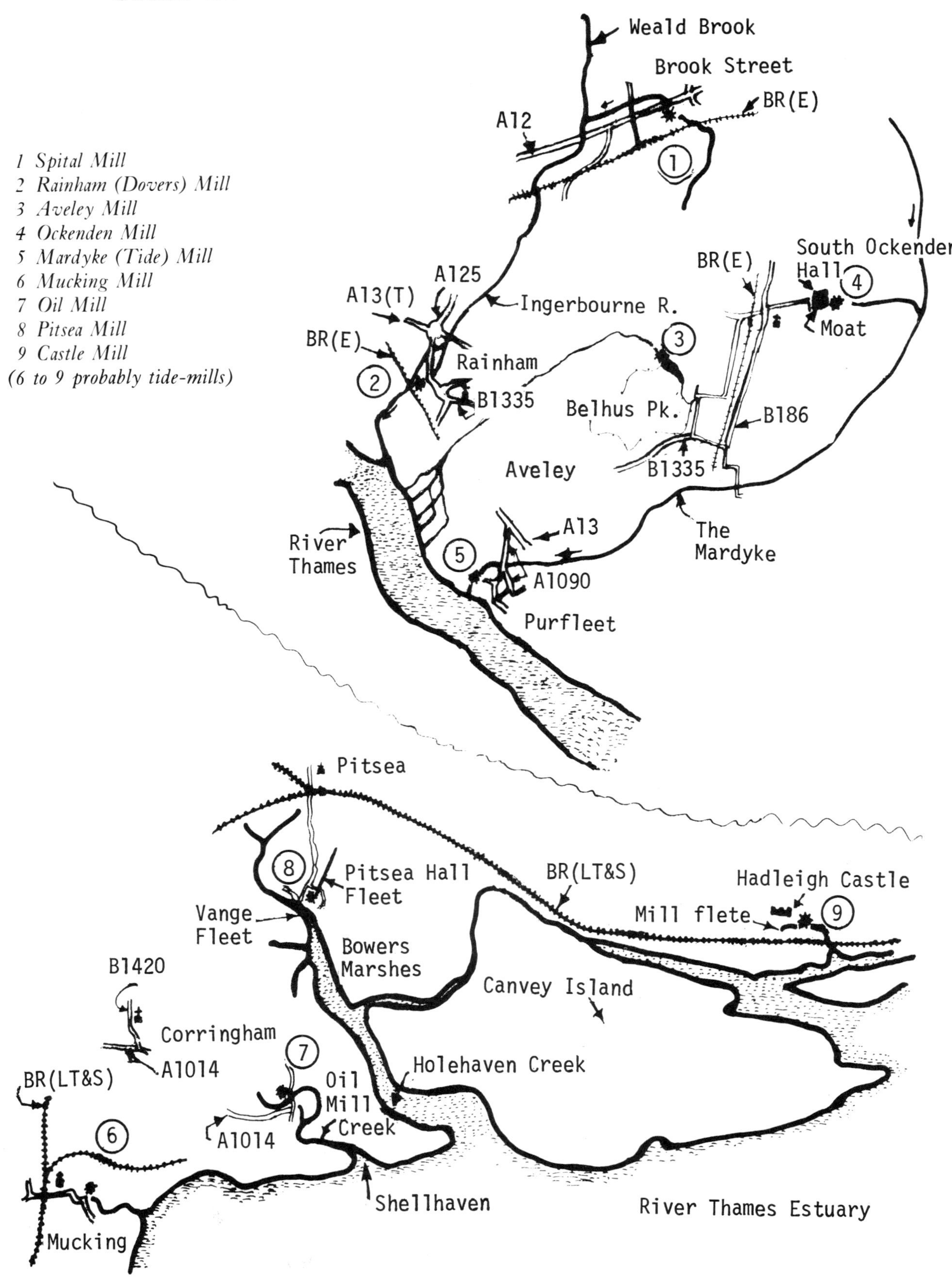

Fambridge Mill, Essex (1952)

Roxwell Mill, Essex (1931)

specimen ground coarse stuff for animal feeding as it does today. An appropriate date for the mill would be 1800, and it is partly brick and weather-boarding with a mansard roof — the lower slope of red tiles, but the upper relaid with blue slates, a cheaper expedient. The adjoining miller's house is seventeenth century, but was re-fronted in the then fashionable manner about the time when the main structure was rebuilt. The scene can be recognized as it appeared forty years ago, but the changes are evident. Of the great water-wheel only the axle, spoke discs, and a few spokes remain; the building is fast becoming derelict, while the loss of the great poplar tree that gave shade and character to this spot, the intrusion of an ungainly concrete ramp, and the less cared-for state of the waterways are conditions, of course, now to be expected. The mill, evidently belonging to Springfield Barnes Manor, was a fulling-mill in the early fourteenth century and was still used for the same purpose in 1442.

Further down the Chelmer, Sandford Mill, typical of the larger Essex watermills, once presented an attractive picture, but such an undeveloped state sealed its fate, for sometime in the early thirties the local water authority removed it for the present waterworks. Further downstream a weir marks the site of a vanished water-wheel, that of the mill attached to Little Baddow Hall, while a mile further along in the same parish stood an old paper-mill. Apart from barns and sheds, nothing survives except sluices and remains of brick work. It has been in existence since 1765 and was the first paper manufactory in Essex. Its predecessor attached in 1294 to Culverts' Manor (half a mile to the north-west) was held by Bishop Burnel of Bath and Wells.

NOTES AND REFERENCES

1 Chapman, J., and André, P., *Map of Essex*, London, 1777
2 Rocque, *Map of London*
3 Randolf, J. A., *Abbeys around London*, Mercantile Press, London, 1899
4 See D. C. Dowsett, *Dunmow through the Ages*, Letchworth Printers, Letchworth, 1971, p. 53
5 Clark, A., *Great Waltham Five Centuries Ago, Essex Review*, vol. XIII, 1904; also Marriage, W., and Smith, C. F., *History of Cornmilling in Essex, Essex Review*, vol. XVI, 1907
6 Apparently now ceasing producing meal

Mill at Barking Creek (1931)

NOTE ON THE MAPS

The maps in the topographical chapters are included for the river basins chosen (i.e. 35 miles from St Paul's). Though drawn from site, they are lettered to enable the sites to be located on ordinary maps.

Sites indicated cannot be equated with surviving watermill buildings; in fact by now, only a very few of those listed will be found to be in such a completed state. The sites represent watermills in a range of conditions, from complete or partial working order down to equally complete disappearance. The list cannot give information as to what condition applies, because so many changes have taken place during the last decade. Many sites are conjectural and cannot be precisely located or even named.

A word as to the names applied to individual watermills: it is a surprising fact that a large proportion bear no individual names, other than those by which the miller or the locals, at the time of asking, happened to know them — let alone their legal names in relevant deeds or records, if the latter chanced to exist. The names set down in the lists are those given locally, or recorded; otherwise the name of the parish, hamlet or manor is used.

These maps are diagrammatic, and not to scale.

POSTSCRIPT

In the foregoing chapters, the telling of what has taken place may appear to have been to the exclusion of what may come about. To redress this unavoidable imbalance, as well as to consider other aspects of the watermill story which have had to be omitted, would need nothing less than an additional volume.

If one had considered such a proposal in the 1960s, however, one might well have confirmed the seemingly inevitable disappearance of the watermill, and have discussed how to reverse effectively this process so that at the least a range of working examples would survive at the end of the decade. The growing demand for wholemeal products would have provided some support, though not overlooking the fact that millstones can be turned equally well, all costs considered, from off the grid, than from a water-wheel. Even further pages would not hesitate to set forth the almost daunting difficulties, technical and administrative, which beset most attempts to retain many mills. This particular objective could not avoid being radically influenced by a greater one, that of the built and natural landscape heritage. In this context, a compromise between applied technology and the surviving environment must be expected, and one that within certain limits might be found to be acceptable, inclusive of curbs both helpful and restrictive.

Another aspect to be debated is that of arriving at a judgement as to how much emphasis should be placed on the watermills' siting, antiquity, or the uniqueness of the plant housed inside. Which should take priority, a fine building having an indifferent complement of internal machinery or one of mediocre visual scoring, yet with an excellent plant? Within an observable consensus that holds at present lies a diversity of opinion. Many maintain that in the plant itself and appurtenances can be found what matters, especially if the latter exhibits the 'standard' kind, or if it illustrates some significant innovation or departure therefrom. Some would recommend the keeping primarily of rare examples, while others think that numbers of the latter, spread countrywide, would perpetuate a false historical perspective. Then there is the claim that early and hence rare specimens of plant, by virtue of that attribute alone, should have an unchallenged priority. Also to be explored, should be the choices of what can and should not be sacrificed in meeting those demands likely to arise in most conversion schemes.

What seemed to be the most successful campaign in this respect would be discussed, namely that in Denmark, where a large number of watermills have been saved, and also taking into account the favourable climate of opinion there, but which did not come forth to accompany those endeavours in the home field.

Other associated matters would be included, such as the disappearance of those craftsmen, millwrights and their kind, whose skills need be available to maintain those examples remaining. Also to be discussed would be vexatious changes such as the imposition of further restriction upon the use of streams and rivers by existing and reinstated water-wheels. This, up to the last war, worked straightforwardly, even in disputes, but enactments since have brought in complications. The streams harnessed then were held to be free and unchallenged for existing wheels, but now such a use has to be paid for, and planning permission obtained. Though the fact of uncertainty can normally be accepted as part of such endeavours, adding more hardly amounts to encouragement, either to keep existing water-wheels turning or reinstating those long since defunct.

The contents would not close without treating of a subject, which though not a structural part, could condition the story. This concerns the nature of the product, at least that which came from most watermills, i.e. wholemeal. Was the original inferior to its later conditioned substitutes? Manorial tenants at most times had to accept what they received rather than what they would have liked. However, a discussion on this subject would soon pass into the land of wide debate, bringing in issues deeper than the mere story of watermills.

Since the 1960s a slow yet persistent enthusiasm has arisen, as witness the number of watermills still in being, some still working and some even flourishing. Hence further achievements in this direction would not be an oversanguine hope. In the relatively short time since then, there have been changes in conviction concerning the more general visual and cultural environment, and these have permeated more than is usually realised. Looking back across the last quarter century, the change does stand out.

The future of the watermill now requires less the need to convince people that those artefacts merit preserving, than in finding appropriate new uses for the redundant structures, or how to ensure that the buildings and plant remain intact from water-wheel to the stones, and waterways to the control sluices, and also with finding the needed resources. Formerly the primary concern was to halt the seemingly wholesale disappearance of watermills, but with this threat lessening, concern has broadened to include making sure that sympathetic expert repair has indeed been enlisted. Not a few of the watermills have in fact lost much in the way of inherited beauty and character, as well as historical interest, but the difficulties now seem to lie more in the realm of logistics, in the identification and availability of the resources which are needed to achieve these worthy objectives.

Town Mill, Watford (1931)

Lucam
Hertingfordbury Mill
(1929)

BIBLIOGRAPHY

Agricola, G., *De Re Metallica*, Ed. H. C. and L. H. Hoover, Mining Magazine, London, 1912.

Ashley, W. J., *Bread of Our Forefathers*, Clarendon Press, Oxford, 1928.

Ashton, J., *History of Bread*, R.T.S., London, 1904

Ballard, A., *Domesday Inquest*, Methuen, London, 1906.

Baring, F. H., *Domesday Tables*, St Catherine's Press, London, 1909.

Beck, T., *Beitrage zur Geschichte des Maschinenbaues*, J. Springer, Berlin, 1900.

Beckmann, J., *History of Inventions and Discoveries*, 4 vols., J. Bell, London, 1897-1914.

Bennett, R. E. and Elton, J., *History of Cornmilling*, 4 vols., Simpkin Marshall, London, 1898-1904.

Berkeley, V., 'Old Worcestershire Watermills', in *Trans. Worcestershire Archarological Society*, Vol. XI New series, 1934, p. 14 *et seq.*, E. Baylis & Sons, Worcester, 1935.

Blith, W., *Six Peeces of Improvement*, J. Wright, 1653.

Bloch, M., 'Avènement et Conquête du Moulin d'Eau', *Annales d'histoire economique et sociale, p. 538 et seq.*, Libraire Arnaud Colin, Paris, 1935.

Boeckler, G. A., *Theatrum Machinarum Novum*, P. Furstein, Nürnburg, 1661.

Bossert, H. T., and Storck, W. F., *Das Mittelalterische Hausbuch*, E. A. Seeman, Leipzig, 1912.

Burstall, A. F., *History of Mechanical Engineering*, Faber & Faber, London, 1963.

Caus, S. de, *Les Raisons des Forces mouvantes etc.*, Frankfurt, 1615.

Camden, W., *Britannia*, G. Bishop and I. Norton, 1610.

Candano, G., *De Rerum Varietate*, Lugdum, 1580.

Clarke, G., *'Watermills in Antiquity', in Antiquity*, Vol. XVIII, no 69, 1944, p. I *et seq.*

Clough, R., *Lead Smelting Mills of the Yorkshire Dales*, Howell & Sowny, Leeds, 1962.

Coleman, D. C., *British Paper Industry*, Clarendon Press, Oxford, 1958.

Cook, T. H., 'Mills of Gade and Colne', in *Hertfordshire Countryside*, Vol. II, no. 43, p. 102.

Coulton, G. C., *The Mediaeval Village*, Cambridge University Press, 1925.

Crozier, M., *An Old Silk Family 1745-1945*, Aberdeen University Press, 1947.

Cunningham, W., *Growth of English Industry and Commerce*, fifth edition, Cambridge University Press, 1915.

Curwen, E. C., 'Problem of Early Water-mills' in *Antiquity*, no. 71, 1944, p. 130 *et seq.*

Dacres, R., *Art of Water Drawing*, Henry Broome, 1660.

Dodd, G., *British Manufactures*, 6 vols., (Charles Knight's Weekly Volumes), London, 1844-6.

Drackman, *Mechanical Technology in Greek and Roman Antiquity*, Munksgaard, Copenhagen, 1963.

Dugdale, W., *History of Imbanking and Drainage etc.*, Alice Warren, London, 1662.

Ferguson, E. S., *Bibliography of the History of Technology*, M.I.T. Press, Cambridge, Mass., 1968.

Finch, W. C., *Watermills and Windmills*, C. W. Daniel, London, 1933.

Fitzherbert, A., *The Boke of Husbandrie*, 1523.

——— *The Boke of Surveyinge*, 1523.

Forbes, R. J., *Studies in Ancient Technologies*, E. J. Brill, Leiden, 1955.

Forest de Belidor, B., *Architecture Hydraulique*, 2 vols., Paris, 1737-53.

Gardner, E. M., 'Some Notes on Dutch Water Mills' in *Newcomen Soc. Trans.*, Vol. XXVII, 1949-51, p. 199 *et seq.*

Garnett, W., *Little Book on Water Supply*, Cambridge University Press, 1922.

Gille, B., *The Renaissance Engineer*, Lund Humphries, 1966.

Gray, A., *The Experienced Millwright*, A. Constable & Co., Edinburgh, 1806.

Gras, N. S. B., *Evolution of the English Corn Market*, Cambridge, Mass., 1915.

Guttman, O., *Monumenta Pulveris Pyrii*, Artists' Press, Balham, London, 1906.

Hogden, M, T., 'Domesday Watermills' in *Antiquity*, Vol. XIII, 1939, p. 261-77.

Hone, N. J., *The Manor and Manorial Records*, Methuen, London, 1906.

Horst, T. van der, and Polley, D., *Theatrum Machinarum Universale*, Amsterdam, 1736-7.

Hudson, H. K., *Handbook for Industrial Archaeology*, J. Baker, London 1967.

Jacob, H. E., *Six Thousand Years of Bread*, E. & C. Winston, Doubleday, Doran, New York, 1944.

Jarrett, H. B., *Social Theories of the Middle Ages*, E. Benn, 1926.

Jenkin, W. R., *The Collected Papers of W. Rhys Jenkins*, Newcomen Society, Cambridge University Press, 1936.

Jensen, L. B., *Man's Food*, Garrards Press, Champaigne, 1953.

Jespersen, A., *Report on Watermills*, Vol. III, published by the author, Virum, Denmark, 1957.

——— *Gearing in Watermills in Western Europe*, Virum, 1953.

——— *Gearing in Watermills*, Virum, 1953.

Landsberg, Harrade de, *Hortus Deliciarum*, Ed., F. C. le Roux, Strasbourg et Paris, 1952.

Laurence, E., *Duty and Office of a Steward etc.*, J. Shuckburg, London, 1727.

Jones, D. H., 'Manx Water Mills' in *Journal of the Manx Museum*, Vol. VII, 1966, p. 114 *et seq.*

Keller, A. G., 'Renaissance Waterworks' in *Endeavour*, Vol. 25, 1966.
Leak, J., *Waterworks*, J. Moxon, 1659.
Leupold, J., *Theatri Machinarum Hydraulicarum*, 2 vols., Leipzig, 1724-5.
—— *Theatrum Machinarum Molarium*, Leipzig, 1735.
Lipson, E., *Economic History of England*, fifth edition, Vol. I, A. & C. Black, London, 1929.
Luckhurst, D., *Monastic Water Mills*, 1963, Booklet no. 8, Society for the Protection of Ancient Buildings, Wind & Watermill Section.
Lyndewode, G., [W. Lyndwood] *Provinciale*, Ed. J. V. Bullard and C. H. Bell, Faith Press, London, 1929.
Maddox, H. A., *Paper, Its History etc.*, Pitman, London, 1916.
Maitland, F. W., *Domesday Book and Beyond*, Cambridge University Press, 1897.
Major, J. K., *Finch Brothers Foundry, Sticklepath, Okehampton, Devon*, David & Charles, Newton Abbot, 1969.
Martini, Francesco Di Giorgio, *Architetto*, Ed. R. Papini, Edn. Electa, Firenze, 1946.
Matthews, W., *Hydraulia, An Historical and Descriptive Account of the Waterworks of London etc.*, 1835.
Messent, C. W., *Old Watermills of Norfolk*, Fletcher & Son, Norwich, 1939.
Miller, W. T. W., *The Watermills of Sheffield*, Pawson & Brailsford, Sheffield, 1949.
Mortimer, J., *The Whole Art of Husbandry*, London, 1707.
Norden, J., *The Surveiors Dialogue*, H. Astley, London, 1607.
Pelham, R. A., *Fulling Mills*, 1956, S.P.A.B. Booklet no. 5.
—— *Old Mills of Southampton (Southampton Papers*, no. 3), Southampton Corporation, 1963.
Quennell, M. & C. H. B., *History of Everyday Things in England*, Batsford, London, 1918-30.
Ramelli, A., *Le Doverse ed Artificiose Machine*, Parigi, 1588.
Rees, A., *Cyclopaedia*, Longman, London, 1819-20.
Robins, F. W., *Story of Water Supply*, Oxford University Press, London, 1946.
Roger, J. E. T., *History of Agriculture and Prices in England*, Clarendon Press, Oxford, 1866-92.
—— *Six Centuries of Work and Wages*, Sonnenschein & Co., London, 1884.
Salzman, L. F., *Building in England down to 1540*, Clarendon Press, Oxford, 1952.
—— *English Industries of the Middle Ages*, Clarendon Press, Oxford, 1925.
Seebohm, F., *English Village Communities*, Longman, 1883.
Shorter, A. H., *Water Paper Mills in England*, 1966, S.P.A.B. Booklet no. 9.
Skilton, C. P., *British Windmills and Watermills*, Collins, London, 1947.
Snape, R. H., *English Monastic Finances, Cambridge Studies in Medieval Life and Thought*, Cambridge University Press, 1926.
Somer-Cocks, H. L. and Boyson, V. F., *Edenbridge*, Edenbridge Chronicle Office, Edenbridge, 1912.
Somervell, J., *Water Power Mills of South Westmorland*, T. Wilson & Son, Kendal, 1926.
Storck, J. and Teague, W. D., *Flour for Man's Bread* [History of Milling], University Press of Minesota, Minneapolis, 1952.
Straker, E., *Wealden Iron*, G. Bell, 1931.
Sutcliffe, J., *Treatise on Canals etc.*, Rochdale, 1816.
Swedenborg, E., *Opera Philosophica et Mineralia* [*De Ferro*], 3 vols., F. Hekelli, Dresden and Leipzig, 1734.
Tann, J., *Gloucestershire Woollen Mills*, David & Charles, Newton Abbot, 1967.
Tawney, R. H. and Power, E., *Studies in Economic and Social History*, London School of Economics.
Tawney, R. H., *et al*, *Select Documents*, G. Bell & Sons, London, 1914.
Telford, W., 'On Mills' in *Newcomen Society Trans.*, Vol. 17, p. 205 *et seq.*
Thrupp, S. L., *A Short History of the Bakers' Company of London*, London, 1933.
Tusser, T., *Five Hundreth pointes of Good Husbandrie etc.*, Assignes of William Seres, London, 1590.
Usher, A. P., *History of Mechanical Invention*, McGraw Hill Book Co., New York, 1929.
Vacandard, E., *Vie de St Bernard*, 2 vols., Paris, 1895.
Vaughan, R., *Long Experience of Water Works, 1610*, Edn. J. Hodges, London, 1897.
Veranzio, F., *Machinae novae etc.*, 5 parts, Venetiis, 1575.
Viollet-le-Duc, E. E., *Dictionnaire Raisonné d'Architecture etc.*, Paris, 1858-68.
Vowles, W. H. P. and W., *Quest for Power*, Chapman & Hall, London, 1931.
Wailes, R., *Tide Mills*, 1956, S.P.A.B. Booklet nos. 2 and 3.
—— *The English Windmill*, Routledge & Kegan Paul, London, 1954.
Wilson, P., *Watermills, An Introduction*, S.P.A.B. Booklet no. 1, 1956.
—— *Watermills with Horizontal Wheels*, S.P.A.B. Booklet no. 7, 1960.
Zonca, V., *Novo Teatro di Machine etc.*, Padua, 1607.

Topographical

Victoria County Histories — Topographical Sections of the counties concerned.

Andrews, J., Herbert, W. and Dury, A., *Survey of Kent*, Dury, London, 1679.
Antrobus, J. J., *Bishop's Hatfield*, Hatfield, 1912.
Ashdown, C. H., *St Albans*, Elliott & Stock, London, 1893.
Ashford, L. J., *History of High Wycombe*, Routledge and Kegan Paul, London, 1960.

Aubrey, J., *Perambulation of Surrey, 1673-92*, E. Curll, London, 1719.

Austin, W., *History of Luton*, 2 vols., County Press, Newport, Isle of Wight, 1928.

——— 'The Domesday Watermills of Bedfordshire' in *Beds. Hist. Rec. Society's* publication, Vol. III, p. 208 *et seq.*, Apsley Guise, 1916.

Becker, M. J., *Rochester Bridge*, Constable & Co., London, 1930.

Besant, W., *Survey of London*, 10 vols., A. & C. Black, London, 1902-12.

Bowen, E., etc., *Large English Atlas*, Carrington Bowles, London, 1785.

Buck, S. and N., *Views*, London, 1726-52.

Burn, J. S., *Henley on Thames*, Longman, London, 1861.

Cary, W., *Survey of Country fifteen miles around London etc.*, London, 1786.

Chapman, J. and André, P., *Map of County of Essex 1777.*

Cobb, J. W., *Two Lectures on the History and Antiquities of Berkhampstead*, S. B. Nichols & Son, London, 1855.

Cobham, D. F., *History and Antiquities of Lingfield*, Cassell, London, 1899.

Crosby, A., *History of the River Fleet*, Guildhall Library MS no 5138, 1845.

Darley, S., *Chapter in the History of Cookham*, privately printed, Edinburgh, 1909.

Dawes, C. E., *Records of Ware*, Geo. Price, Ware, 1901.

Defoe, D., *Tour through England and Wales, 1727*, Everyman's Library no. 821, J. M. Dent, London, 1928.

Derry, D. K. and William, T. I., *Short History of Technology*, Clarendon Press, Oxford, 1960.

Dews, N., *History of Deptford*, J. D. Smith, Deptford, 1883.

Dudley, H., *Description of Principal Towns in Sussex*, Eastbourne, 1835.

Duncan, L. L., 'History of the Borough of Lewisham' in *Proceedings of Lewisham Antiquarian Society, 1908-12*, Lewisham Council, Lewisham, 1963.

Dunkin, J., *History and Antiquities of Dartford*, Dartford, 1844.

Eland, G. E., *Old Works and Past Days in Rural Buckinghamshire*, G. T. de Fraine & Co., Aylesbury, 1921.

Ewings, G., *Westmill*, Courier, Tunbridge Wells, 1928.

——— *History of Cowden*, Courier, Tunbridge Wells, 1926.

Finch, W. C., *The Medway, River and Valley*, C. W. Daniel, London, 1929.

——— *In Kentish Pilgrim Land*, C. W. Daniel, London, 1925.

Fisher, J. L., *The Deanery of Harlow*, Benham & Sons, Colchester, 1922.

Ford, E., *History of Enfield*, Enfield, 1873.

Fox, W. H., 'Wymondley Priory' in *East Hertfordshire Archaeological Society*, The Camden Press, 1903.

Fry, K., *History of East and West Ham*, privately published, London, 1888.

Gardner, E. M., *Three Mills, Bromley-by-Bow*, 1957, S.P.A.B. Booklet no. 4.

Goodsall, R. H., Watermills on the River Len, in *Archaeologia Cantiana*, Vol. LXXI, p. 106, Headley Bros., Ashford, 1958.

Greenwood, C., *Map of Kent*, 1821.

——— *Map of Middlesex*, 1819.

Guiseppi, Montague S., 'The River Wandle, Past and Present', in *The Wimbledon & Merton Annual*, 1910.

Gyll, G. W. I., *History of Wraysbury etc.*, London, 1862.

Hale, W. H., *Domesday of St Pauls, London, 1858*, Camden Society Publication no. LXIX.

Harben, H. A., *Dictionary of London*, Jenkins, 1918.

Hardy, W. J. (Ed.), *Home Counties Magazine*, Vol. 1-14, London, 1899-1912.

Harris, J., *History of Kent*, D. Midwinter, London, 1719.

Harwood, T. E., *Windsor Old and New*, privately printed, London, 1929.

Hasted, J., *History of Kent*, Simmon & Kirkby, Canterbury, 1778-99.

Heales, A. C., *Records of Merton Priory*, H. Frowde, London, 1898.

Hillier, J. R., *Old Surrey Water Mills*, Skeffington & Sons, London, 1951.

Hine, R., *History of Hitchin*, 2 vols., Allen & Unwin, London, 1927-9.

Hope, W. H. St John, *Windsor Castle*, 3 vols., Country Life, London, 1913.

Hopkins, R. T., *Old Watermills and Windmills*, P. Allan, London, 1930.

——— *Old Mills and Inns*, Palmer, 1927.

Horsburgh, E. L. S., *Bromley, Kent etc.*, published by the Bromley Committee, 1929.

Horsfield, T. W., *History etc. of the County of Sussex*, 2 vols., Sussex Press, Lewes, 1835.

Ireland, S., *Picturesque Views on the River Thames*, 2 vols., London, 1792.

James, N. G. Brett, *Growth of Stuart London*, London & Middlesex Archaeological Society publication, 1935.

Johnson, W. B., *Industrial Archaeology of Hertfordshire*, David & Charles, Newton Abbot, 1966.

Kerry, C., *History and Antiquities of Bray*, privately printed, London, 1861.

Lathbury, R. H., *History of Denham*, published by the author, Uxbridge, 1904.

Lyson, D., *Environs of London*, 4 vols., London, 1792-6.

Maidstone, *Topography of Maidstone and Environment etc.*, Maidstone, 1839.

Major, J. K., 'Berkshire Watermills' in *Berkshire Archaeological Journal*, Vol. 61, pp. 83-91, 1963-4.
Mayes, L. J., *History of the Borough of High Wycombe*, Routledge & Kegan Paul, London, 1966.
Mitton, G. E. (Ed.), *London Maps and Drawings*, A. & C. Black, London, 1908.
Morant, P., *History and Antiquities of Essex*, London, 1768.
Mudge, W., *Map of Kent*, Faden, London, 1801.
Newton, W., *History and Antiquities of Maidstone*, London, 1741.
Norden, J., *Description of Essex*, Camden Society, Vol. IX, London, 1840.
Ogilby, J., *Britannica*, A. Swall & R. Morden, London, 1698.
Pearce, E. H., *Monks of Westminster*, [Notes and documents relating to Westminster Abbey, no. 5], London, 1909.
Pearson, H., *Memorials of the Church and Parish of Sonning*, E. S. Blackwell, Reading, 1890.
Phipps, P. W., *Chalfont St Giles, Past and Present*, R. Bentley and Sons, London, 1896.
Pinks, W. J., *History of Clerkenwell*, London, 1865.
Redford, G. and Riches, T. H., *History of Uxbridge*, W. Lake, Uxbridge, 1818.
Robinson, W., *History of Hackney*, 2 vols., J. B. Nicholls & Sons, London, 1842.
——— *History of Tottenham*, Tottenham, 1818.
Robo, E., *Mediaeval Farnham*, E. W. Langham, Farnham, 1935.
Rocque, J., *Map of London*, 1746.
——— *Map of Middlesex*, 1754.
Royal Commission on Historic Monuments — volumes on Buckinghamshire, Essex, Hertfordshire and Middlesex.
Russell, J. M., *History of Maidstone*, W. S. Visish, Maidstone, 1881.
Salmon, J., *Guide to Sevenoaks*, J. Salmon, Sevenoaks, 1897.
Sharpe, M., *Antiquities of Middlesex*, 8 parts, Brentford Printing and Publishing Co., Brentford, 1905-16.
Sheahan, I., *History and Topography of Buckinghamshire*, London, 1862.
Smetham, H., *History of Strood*, Parnell and Neves, Chatham, 1899.
Spain, R. J., 'Len Watermills' in *Archaeologia Cantiana*, Vol. LXXXII, Headley Brothers, Ashford, Kent, 1968.
Speed, J., *Theatre of Great Britain*, J. Sudbury & G. Humble, London, 1611.
Stow, J., *The Survey of London*, Everyman Edition, J. M. Dent, London, 1912.
Sutton, C. N., *A Short History of Withyham and Buckhurst*, A. C. Balwin, Tunbridge Wells, 1893.
Thacker, F., *The Thames Highway*, F. S. Thacker, London, 1920.
Thorne, J., *Handbook to the Environs of London*, J. Murray, London, 1876.
Thornbury, W. and Walford, E., *Old and New London*, 8 vols., Cassell, London, 1897-8.
Tregelles, J. A., *History of Hoddesdon*, Austin & Sons, Hertford, 1908.
Turnor, L., *History of Hertford*, Hertford, 1830.
Walford, E., *Old and New London*, 6 vols., Cassell, London, 1893.
——— *Greater London*, 2 vols., Cassell, London, 1890-5.
Westlake, H. F., *New Guide to Westminster Abbey*, A. R. Mowbray, 1916.
Wheeler, L. S., *Chertsey Abbey*, Wells Gardner, London, 1905.
White, C. H. E., 'Church and Parish of Chesham Bois, Bucks.' in *Records of Buckinghamshire*, Vol. 6, p. 179-211, 1886-70.
White, W., *History, Gazeteer etc. of Essex*, R. Leader, Sheffield, 1848.
Whiteman, R. J., *Hexton, A Parish Survey*, E. Cummins, Bradford, 1937.
Wilson, Aubrey, *London's Industrial Heritage*, David & Charles, Newton Abbot, 1967.
Wooler, W. U., *A Short History of the Parish of West Wycombe*, Buckinghamshire Free Press, High Wycombe, 1925.

Rear of Birchen Bridge Mill

(destroyed between 1931-63)

INDEX TO VOLUMES I AND II

Entries in italics refer to Illustrations

C

M

Q

R

T

X

Y

Z